MAKERS BIBLE

A MANIFESTO FOR
QUALITY, PASSION & CRAFTSMANSHIP

INTRODUCTION

Mankind started on his path to separation from the beasts by gathering around a fire and cooking food. Communication as a tribe naturally followed. This enduring tradition can still be seen today in rural homes around the world where families of three generations or more may gather together in the kitchen. In the 21st century, these core ideas are being rekindled by creators of urban living who incorporate modern, open kitchen architecture.

These passionate Makers and their brands represent products that are meant to last and to be passed on to the next generation. Utensils, tools and goods that are as necessary to collecting stories and memories as they are to preparing food, hunting, or fishing. The foods curated here are only a fraction of what people might want to bring to the table.

Environment and society have a great influence on products. The authors of this book present these Makers and their products through a unique cultural filter. The tradition of food, from provision to preparation through consumption, represents whole subsets of our oldest crafts as a species. Man had to eat first, feeding on what was available in the surroundings before anything else. Hence the crafts revolving around food and providing shelter are the oldest there are.

The makers in this book range from traditional Makers to the start-ups of the 21st century with all variants of manpower. What they all have in common is a desire to craft, to work with one's hands, in as many steps of the process as possible. Maintaining or enhancing a flavor, solid or liquid; preserving; seasoning; cutting; grilling; heating; all of these steps are encompassed by this philosophy of craft.

The 101 brands complement each other. Just like the hearth fire and the steaming pot hanging over it. This would not have been accomplished without the brands' contribution. The Makers Bible Beetroot & Steel Edition is a manifesto of passionate manufacturing brands of all levels of craftsmanship but what they have in common is that they enable consumers to become Makers themselves – makers of memorable moments shared with friends.

Beetroot & Steel stands for true and honest products. For the fact that everything one provides for oneself, whether it be from the forest, the river, the sea, the garden or from a merchant and brand one trusts, is authentic, timeless and free of doubt. A freedom that enables one to provide undivided attention to the family and friends gathered around the table.

Sincerely,
The Editorial Team

MAKERS BIBLE
BRAND
COUNTRY
NEARBY

STORE FINDER

www.makersbible.com

The online Makers Bible store finder enables you to convince yourself by discovering the products displayed in this book "in the flesh", trying them on, feeling their weight, exploring their details, tasting them.

In short: Go out and find them. Most of them are not available on every corner. We wanted to keep the process as simple as possible. All you need to do is decide whether you want to look up a brand and find the nearest distributor or even the maker himself. Or maybe you want to plan a trip to a certain city for shopping? Simply type in the city's name and a map or list will display what you are looking for. The store finder is web-based. There is no need to download any app.

We would like to enhance and personalize your makers bible experience so you can also see and hear the testimonies and experience from our team of Makers directly at www.makersbible.com

CONTENTS

AMBASSADORS

BEE HIVE YOURSELF

A TALE OF BEES AND HONEY

From as far back as 40 million years before the evolution of man, honey has been the oldest sweetener in existence, but honey, in ancient history, was not just used as a sweetener. It could be found as an ingredient in many other products such as beverages, medicines, even cement. There are records of lords being paid with beeswax and honey by peasants in Germany, proving that honey was seen as an equivalent to money across various cultures in the old world. Honey has pride of place in nearly every civilization, globally. Indispensable in our kitchens and 20 percent sweeter than refined sugar, this liquid gold of the bees can add a caramel-malty flavor to any recipe.

SCHMIDT & BRIONGOS

SCHMIDT BRIONGOS

The most important benefit of bees, however, is not their honey, but the job they do pollinating our nature, and they are pretty underpaid at an estimated economic value of 70 billion dollars worldwide. These days the bee has become dependent on the aid and care of a beekeeper. Unfortunately, over the last couple of decades, cheap, imported honey has caused a decline in prices and is one of the reasons fewer people are keeping bees. With the ease of import driving the local bee populations down, pollination, a bee's main job, cannot be effectively completed in many European and North American regions.

EVERY THIRD BITE WE EAT DEPENDS ON BEES

This has begun creating devastating consequences for agriculture and biodiversity alike. Over 80% of domestic crops and wild plants are dependent on cross-pollination. This means every third bite we eat is conditional to the existence of bees. The decrease of local bee populations will not only cause a general decline in the biodiversity of nature but will affect a large number of fruits, berries, vegetables and nuts - foods we will desperately miss in our menus and on our plates.

ONE OF THE MOST URGENT PROBLEMS OF OUR TIME

But there is still more going against the bee today. Climate change, invasive parasites and the increasing use of pesticides have led to a decline of bees worldwide. As a result, global initiatives and grass roots movements have begun to emerge to preserve bees and raise awareness for the importance of a bee-friendly environment. One of these initiatives is nearBees - Next Door Honey, a German based start-up (www.nearbees.de).

"With nearBees we aim to fight against the shrinking bee population as one of the most urgent problems of our time.", Viktoria, one of the founders, explains.

Keeping bees has a long tradition in her family. Having fostered colonies for more than six years, she is upholding this tradition. Confronted with the labor-intensive selling of her own honey, she started to analyze the changes in coexistence between mankind and the bee. The outcome was nearBees, an uncomplicated market access for (hobbyist) beekeepers to sell their honey, finance their hobby and maintain keeping bees - both offline and online.

HONOR LOCAL

Without time-consuming research, honey lovers can seek local product right in their neighborhood and directly purchase honey from a nearby beekeeper. To make honey distribution financially attractive they invented a new packaging: the nearBees honey pouch. Packed up flat in an envelope the honey is sent via mail, with a minimum of shipping costs and material consumption.

"By only sourcing local honey, everybody can do their part to stop the decline of bees. For more beekeepers, more bees and the preservation of a blooming and beautiful nature!" Viktoria describes the positive impact of local honey consumption.

Ruben, a beekeeper from Spain, shares the unique story of his personal mission.

" I came to learn that the bees we have now aren't the same size as they were originally in nature. They have been artificially enlarged by humans to be more productive, with larger stomachs to store more nectar. But this has also made them less resistant, especially to the Varroa, a mite that is deadly to large bees, but has a negligible effect on smaller ones. Bringing bees back down to their original size is not only possible, but should also be relatively straightforward, given enough time and patience. And that's what I am trying to achieve."

SCHMIDT & BRIONGOS

BEES SHOULD BE SMALLER AND THEREFORE HEALTHIER

European and North American bees cannot deal with the Varroa, and Ruben intends to change that. It is also critical, as Ruben points out, to buy real honey. With today's imports and artificial product technologies, additives can allow for the creation of 'food products' that resemble but in many ways do not come near the food we think we are buying.

"Here in Spain you can read in the labels: From and not from the EU. Meaning that almost 80% of our honey comes from China or Argentina. Sometimes it is even some kind of syrup made with soya flour. When you heat honey over 40°C you spoil fructose and glucose into saccharose. Supermarket honeys that always maintain a liquid form are heated to over 100°C. Also, usually filtered with such a small filter that it doesn't contain any trace of pollen. So, you couldn't say where this honey comes from. You cannot even call it honey if it doesn't have pollen in it."

TELL YOUR CHILDREN THE TRUTH ABOUT BEES AND BLOSSOMS

Both of these projects share the common goal of spreading awareness of the importance of bees and educating customers on the double impact they can have by buying local honey. By buying local, you become an ambassador for the presence of bees in your region – bees that carry out essential pollination and produce rich tasting, healthy and natural raw honey.

A KNIFE A COFFEE MACHINE A COOKBOOK

KITCHEN DESIGN AND SOCIAL RESONANCE

The kitchen is a location for tragedy and for comedy. A place for resonance. We spoke to Marc O. Eckert owner of bulthaup about kitchens, design and social resonance.

MARC O. ECKERT

Bulthaup has created kitchens in lower Bavaria since 1949, each one born in their workplace in Aich, near Bodenkirchen. There we met a dedicated team of passionate Makers who seem to have kitchens on their minds for far longer than regular working hours.

With its first furniture, back in 1951 bulthaup re-invented the idea of where we prepare and eat food, thinking very differently than most other household supply companies, food producers and furniture carpenters. One may recall those nostalgic food commercials from the 1950s and 60s that portrayed a lonely serving woman in a separate room of a newly built house or flat. Today we look at those messages as an amusing reminiscence of times when roles for people, and the rooms they dwelt in, were clear - but limited.

KITCHENS ARE NOT SOLELY FOR COOKING - BUT FOR COMPANY

In the 1980s bulthaup started to work with German designer Otl Aicher who defined the kitchen as a place "inviting enough to create food together, that should hold kitchen utensils in reach and sight". Not really a new idea, when one thinks of farmhouses where the kitchen always was the place to create food and consume it, but in the 1970s and 80s this had fallen out of fashion in the west.

As the leading kitchen manufacturer in Germany, we found that Mr. Eckert, the third generation of his family to take the helm of bulthaup, had surprising views on what's important in a kitchen. Speaking little about the merits of the kitchen as a whole, he focused mostly on the secondary items found in it.

A kitchen must provide room for a table. A place to gather all members of the tribe. May it be a family of blood or choice, and the friends. No family member should be obliged to run in and out from another

MARC O. ECKERT

MARC O. ECKERT

room and hence miss the conversation. The good and modern kitchen keeps the narrative active by not distracting too much, enabling one to find needed items in an instant and leaving enough time for the most important social attributes of intimacy and reflection.

A bulthaup kitchen is not a practical workstation for professional chefs. It is a crafted assortment of pieces that, in combination, fulfill a transcultural need: The need to gather and make this gathering a potent experience. From a quiet one-on-one, to a tragedy or comedy, savoring good flavors and stimulating connections over a shared pleasure to eat and drink well, fulfills the power-of-place.

FIRE, WATER, FOOD & COMMUNITY

At the core of bulthaup's set of values are four essential needs. The first one is fire, representing not only the cooking of food but also warmth and belonging. Where there is fire, there is a place to appreciate nearness. Second comes water. A most important, life-sustaining element. When one is hydrated one feels well. Having water is a luxury and a means of survival. Third there is food, but surprisingly not the edible kind. In the context of bulthaup it is the soul-food, of resonance and feedback, the food for one's thoughts. The fourth essential is community, a consistent transcultural need, the essence of which can be found in all bulthaup made kitchens.

These core values are often subliminally impressed into the environment they deliver to their users. What is more obviously present to the senses and most often received as carpentry work in perfection, is the craftsmanship and quality standards taken most seriously by the employees of bulthaup.

On our visit, we found one work station empty. The man's task working there was to check the flawlessness of laminate. He does it with his eyes detecting even the smallest imperfections, should they occur, and with both hands. One hand using the naked fingers

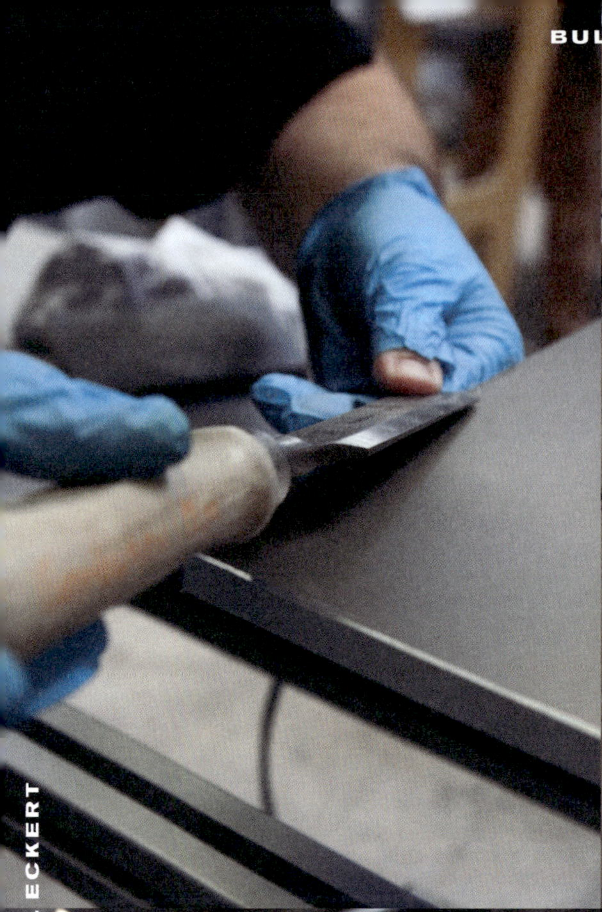

MARC O. ECKERT

MARC O. ECKERT

MARC O. ECKERT

and one using a glasèe glove. The station was empty since he needs to take regular breaks in order to reliably execute his job to perfection for four hours at a time. The fingers and eyes need a break. At bulthaup many tasks are rotated between the multi-skilled workers.

OUR FINGERTIPS TELL FAKE FROM AUTHENTIC, WELL FROM BAD

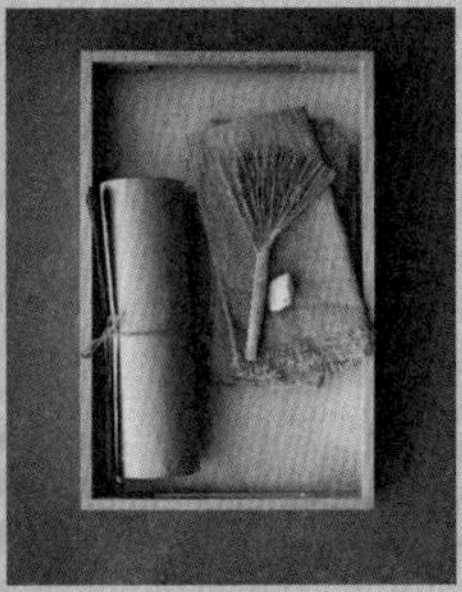

bulthaup uses the term "truth of material" (Materialechtheit) a lot. Even if a surface is refined to perfection, one should still be able to experience the material viscerally, by smell, sound and touch. It is that moment when the tactile senses tell us that a thing is authentic or fake, treated well or bad. And this is the same story for many products in this book and our previous book – a truth that separates real craftsmanship from simply building something.

Superior design is the result of a function or functions that need to be achieved in a certain context and the abilities of the craft and technology at hand. Of course, bulthaup's workshop is not a carpenter workshop with lots of chipped wood and dust floating in the air. It looks rather high-tech, machines doing the rough part. But at bulthaup's core are skilled hands and brains bent toward a single cause, building kitchens that last, by stability and design. To quote Mr. Eckert: "bulthaup starts where the machine ends."

Mr. Eckert spoke to us for a bit about the company's ability to design thin finishing veneer, of horizontal lines, of wood textures and the combination of new and traditional surface materials. He surprised us again when we asked what he looks most for when entering a kitchen. "The character of a kitchen, of what makes 'a kitchen' to my mind, are three items. First are the knives. A blade that has been sharpened over time and now reflects the owner's way of use. Second is the coffee maker. Not the hole in the wall or the Nespresso machine but the Espresso machine or simply the 'Mocca'. And third the book or books. A cook book or a note book with dog-ears, fat stains and flour in its folds."

Our conversation brought us inexorably back to the foundational purpose of a bulthaup kitchen. Of all the attributes, that are optional choices in a bulthaup kitchen, like the type of wood, of texture, of surface colors, of product options and interiors, of drawers and cabinets ... it comes down to feeling well in it.

Eckert explained, "If I feel well and can make others feel well in it, in the kitchen and in my company, does the quality of the food matter too much? Not really." He continued, "It's about the stance. Not the brand."

Craftsmanship to fulfill a desire. All craftsmanship at bulthaup is not only a matter of prestige, even if there are customers of bulthaup that buy a bulthaup because of the name, but it is driven to create a space for the feeling of belongingness and the joy of having a good time.

bulthaup considers their use of external designers as "intelligence amplifiers". Recognizing the need for innovation through iteration and the power of injecting new thought patterns into the mix, they partner with these intelligence amplifiers and essential ideas are brought in. These ideas, however, are developed in teams inside the company and cross-departmentally and simultaneously with other external partners such as mechanical engineers who customize machines to their needs and more so to their ideas.

Modifying tools to fit their needs and bringing new thought patterns and skillsets in to achieve whatever singular vision is necessary for a given kitchen environment, typifies the bulthaup way of bringing to bear everything essential to create a power-of-place, embodied with their core values. bulthaup doesn't just connect a group of furniture to some countertops. They assemble environments uniquely suited to food preparation, consumption and all of the social interactions bundled around those that the human being requires.

MARC O. ECKERT

MASTER OF BLACK MAGIC

EDY BIEKER EVERTHING ABOUT COFFEE

Producing a coffee requires two kinds of handcraft: first, in the roasting and second, in the liquefying and pouring it into a cup. Italy seems to be the only culture that has mastered these crafts to perfection: to produce a decent espresso on every corner of the country. Edy Bieker, head of quality and training at Sandali Trading Company S.p.A., Italy's largest single coffee importer, based in Trieste ("città di caffè"), on all things coffee.

EDY RIEKER

EDY BIEKER

Coffee and espresso are served around the world. Baristas of every creed have been claiming fame for concocting hosts of outstanding well-prepared, caffeinated beverages. These local champions in coffee can be found in little towns and big cities from the US, New Zealand, the UK, Spain, Austria, Switzerland and Japan, just to name a few.

ITALY'S GIFT TO THE WORLD WAS NOT THE INVENTION BUT THE PERFECTION

But Italy is one of the few nations that has embedded coffee in its culture so deeply that one can almost be 100% certain that if the desire for a coffee arises, any road side café, bar, coffee shop or restaurant from the Brennero in the north to Palermo in the south, from Milano to Napoli that there will be that small quantity of the black liquid with a bit of cremá on top, always available. Italians are passionate about their coffee. While they didn't invent it, they have birthed a coffee culture unequalled anywhere else in the world.

EDY BIEKER

We asked world-renowned coffee expert Edy Bieker, winner of the SCAE Lifetime Achievement Award and head of quality and training at Sandalj Trading's. The lead to Mr Bieker was provided by Procaffè. Procaffè is not only one of Italy's largest coffee roasters but also owner of brand Bristot (p. 84) and Vescovi (p.124). Edy Bieker consults both brands on their quest in seeking the perfect coffee taste. He has had a love affair of over four decades with this sultry brown liquid, "Why Italy?" His answers and knowledge about the bean were eye opening.

"When it comes to coffee, there are always two cooks involved.", Edy explained. "The first cook is the roaster: a chef that selects the product, cooks it and enables it to mature properly. They transform the coffee bean using alchemy. The second chef is the person that uses the machine to extract the coffee, cooking it again, transforming it from a solid to a liquid. Both are very delicate phases. Once selected, raw coffee can never be improved but it can certainly be ruined in the subsequent processes if they are not completed with the necessary accuracy."

Espresso, we learned, was Italy's gift to the world. Italians discovered a quick alchemical method that had never before been seen. It transforms a solid into a liquid via a 25-second extraction process, resulting in a reduced waiting time between the product being freshly ground and the coffee ending up in the cup. It is a kind of chemical process that transforms coffee powder into a black liquid with a crema on top. Other extraction methods, such as filtration, generate a lighter liquid drink. Italian espresso is an extract whereby a concentrate is generated through a process of sublimation.

When the coffee used is of good quality, this method generates a good espresso. However, when the coffee is of a bad quality, it risks bringing out the negative characteristics more than when using the filtration method. This is why espresso is such a refined, delicate product. Espresso is a true measure of how high the quality of the growth, roast and grind for the bean is.

The most important part of an espresso is the crema. An absence of crema can be quite jarring to the taste. Even if the casual drinker does not fully understand the sensory depth of a coffee, they can still enjoy the color and consistency of the crema in their cup. The success of the crema can depend on many factors. The amount of coffee used, the temperature of the water, the age of the product and the amount of time that has passed since the coffee was extracted are the most telling. Simply having a good blend is not enough – espresso is an art as well as a science.

"Coffee still surprises me. One can still find poor-quality products on the market." Edy maintained. "Our years of experience have seen us raise the bar in terms of quality time and again. We look at the aromas in our product and figure out the winning combination between body and persistence and aromatic quality,

EDY BIEKER

but this will always be a combination that is open to interpretation and evolution through taste."

The professional coffee tasters try to check whether there are any evident issues with the coffee extraction process. At the same time, they are looking for the product characteristics they hope to find when they buy a coffee. The sensory quality must live up their expectations.

To ensure that a coffee has no defects, one has to be an expert in selecting coffee, an expert in roasting and an expert in extraction. That makes for a perfect product. At that point, subjectivity comes into it. One experiences different ways of brewing or extraction and the various aromatic notes in a coffee, enjoying them in different ways.

The majority of Italian households' drink coffee prepared in a moka. Americans, Germans, Dutch percolate or brew coffee. The moka pot is a direct consequence of espresso. It's another quick way of extracting coffee. The two extraction methods and the percolation process all result in different sensory experiences, but no one is necessarily better or worse than any other. They're just different. It's just like eating meat roasted, stewed or boiled. The concept is the same.

There are almost as many different ways to grow, roast, grind, prepare and serve coffee as there are peoples spanning this earth. But if we've learned anything from our foray into the Italian espresso alchemist's world of tasters, and extractions, we've learned that one can always find a common ground with a perfect espresso in Italy.

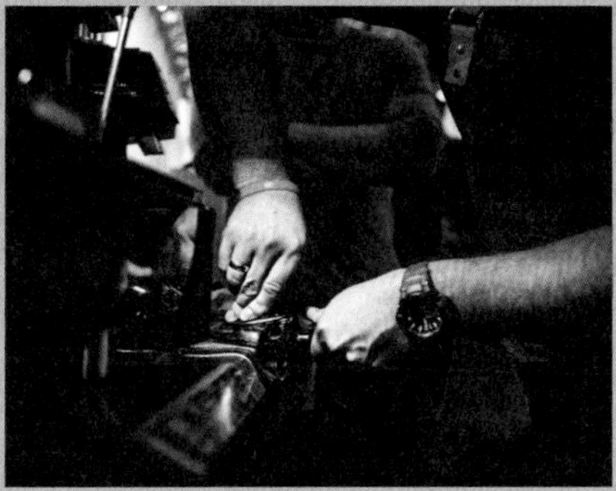

HERITAGE ON THE MENU

140 YEARS OF KITCHEN CLASSICS

Since German home appliances manufacturer NEFF pioneered the first stove and oven in 1877, food has changed a lot, shaping the way our kitchens and homes look, and how we interact with friends and family. The evolution of food speaks to our evolution as individuals, as families and as nations. To celebrate 140 years, NEFF has partnered with Olia Hercules, a Ukrainian born food anthropologist, to reveal a timeline of dishes and recipes influenced by social and political changes in Europe over the past 14 decades.

OLIA HERCULES

The resulting narrative of our recent food history brings these trends and NEFF's rich history to life as the two interweave throughout the decades, guiding what and how we eat today. Born in Soviet Ukraine, with a family history that covers half of Eurasia, Olia Hercules has always been passionate about food and has spent plenty of time researching cuisines. Her family left the Ukraine when she was 12, emigrating to Cyprus, and she continued to discover more of Europe's countries and culture as her studies took her first to the UK and then to Italy to complete her Master's degree. Olia decided to settle in London, where she pursued her dream of cooking for a living. Author of the bestselling cookbook Mamushka, which celebrates her family recipes, she launched another book in 2017 called Kaukasis: The Cookbook – A Journey Through the Wild East.

2. Mini Churros
Spain 1887-1897
Mini churros paired with chocolate sauce are a delicious representation of this era and a still booming phenomenon.

Olia's 14 decades of Dinner research takes inspiration from key food trends throughout Europe over the past 140 years. Each decade is represented by an iconic dish inspired by the historical developments and transitions occurring in the dish's country of origin. The resulting feast showcases an inspiring display of Europe's culinary breadth and celebrates each country's contribution to today's multicultural tapestry of food.

1. HUTZELBROT

Germany 1877-1887

At the start of our 140-year journey, the industrial revolution was in full swing. The result was a rapidly emerging working class with a demand for faster, nutritious meals to feed them. One solution was delicious and nutritious Hutzelbrot, made with dried fruits, nuts and pears. At the same time, in 1877, a flame was lit and the first roast was cooked in a coal oven from NEFF.

1

11
3
6

3. MELBA TOAST CANAPÉ

France 1897-1907

This decade is represented by Melba toast canapé, topped with pigeon and chestnut pâté is in this combination, inspired by an almost forgotten Viennese dish from the era. Auguste Escoffier, the French chef, restaurateur and culinary writer, heavily influenced this decade. Escoffier popularized and updated traditional French cooking methods, significantly impacting the food culture in both France and Britain. His recipes and approaches to kitchen management remain influential today. In 1897, following the illness of famed opera singer Dame Nellie Melba, during which she was only able to eat toast, Escoffier invented a dry, crisp and thinly sliced toast in her honor, which was subsequently christened 'Melba Toast'. This decade also saw the rise of cafés as a new type of public forum, as the bohemians called bourgeois norms into question with their anti-establishment and unorthodox lifestyle. Carl Neff was quick to recognise that confectioners needed better ovens if they were to satisfy the sweet cravings of these customers who regularly frequented cafés, paving the way for the invention of gas burners for bakeries.

4. OXO cube inspired broth
Britain 1907-1917

5. Buck's Fizz
Britain 1917-1927

8. BISTECCA ALLA FIORENTINA

Italy 1947-1957

Bistecca alla Fiorentina is a favorite of Tuscan cuisine in which a Florentine T-Bone steak is marinated in balsamic vinegar, anchovies and herbs. During the aftermath of the Second World War, Italy became prosperous and the nation saw increasing numbers of women enter the workforce. Consequently, they valued the precious time they had outside of the workplace more and began to abandon recipes that took a long time to prepare, in favor of simpler and quicker dishes. In Germany, out of the rubble of post-war emerged the economic miracle. "Made in Germany" became the mark of quality for export goods and NEFF was riding this wave of success. The demand both at home and abroad grew exponentially, as did the number of employees. In 1950, NEFF became the first manufacturer to install thermostats in standard ovens.

6. Smorgasbord
Sweden 1927-1937
11. Endives canapé
Belgium 1977-1987

8

10

13

7. Woolton Pie
Britain 1937-1947

10. PICKLES

Eastern Europe 1967-1977

While other parts of the world enjoyed rapid economic growth, people all over the Soviet Union were experiencing food shortages. Empty shelves, which often only held bottles of vinegar, meant that pickling was on a massive rise as the nation found ways to preserve the food available to them. Elsewhere in Europe, food preservation was also high on the agenda, "a fridge in every home" became the symbol of the new Zeitgeist.

13. GRILLED OCTOPUS

Spain/ Greece 1997-2007

Grilled octopus with skordalia, a garlicky mashed potato purée with pickled rhubarb and radishes, highlights Spain and Greece, the original 'sharing' nations. This is the decade that kick starts the beginning of the sharing plates era.

14. BLACK FOREST GATEAU

Germany 2007-2017

To celebrate 140 years of NEFF, its role over the past 14 decades and the production of its millionth Slide&Hide® oven, the focus is on a modern interpretation of the Black Forest Gateau, nodding to NEFF's German origins.

The journey from 1877 to 2017 reveals a rich tapestry of trends, which interweave throughout the decades, ebbing and flowing as the continent's relationship with food evolved. As present technology continues to advance in kitchen innovation (see a few of NEFFs contemporary developments on page 176), we hope that the kitchen becomes a place of refuge, where foodies can enjoy the simple pleasures that preparing food for friends and families can provide.

9. Cheese Fondue
Switzerland 1957-1967
12. Sourdough Bread
Italy 1987-1997

OLIA HERCULES

(14)

PRAISE THE LARD

MATTEO ZANUS

NOT YOUR AVERAGE ITALIAN CHEF

Cook Free - tattooed across eight of his knuckles says it all: This Italian chef from Veneto is anything but tomatoes, mozzarella and basil. Teochef is a place of originality and attitude. Like it or not. And even if you come to the conclusion his place, right by the Ponte Vecchio in Bassano del Grappa, is only about meat, don't be fooled, Matteo Zanus will strike again!

MATTEO ZANUS

MATTEO ZANUS

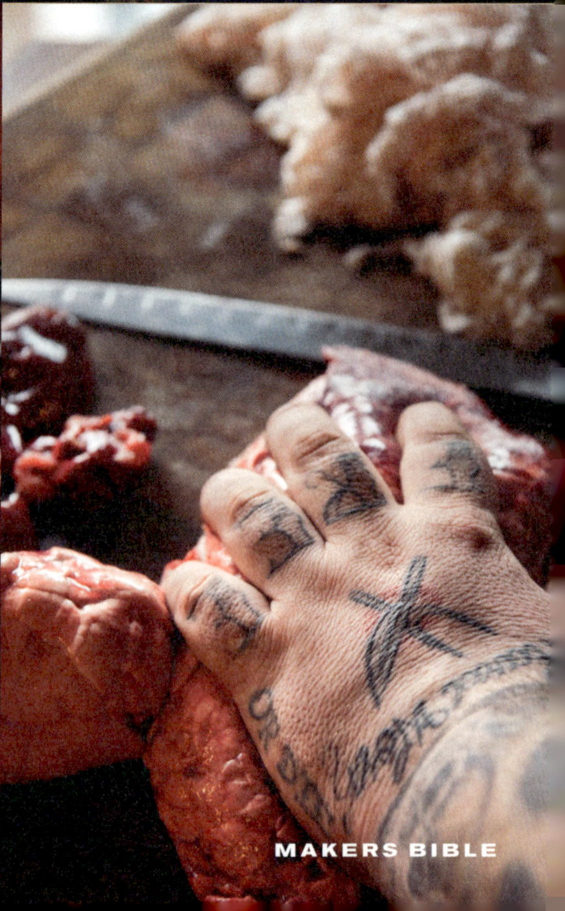

THE MAN:

Cook Free or Die, emblazoned across the tops of his hands, reminding him every day, not to follow the masses and trends but to be unique, to do his own thing. A cooking philosophy that Matteo Zanus Aka "Teochef", owner and EL Capo of Cucina Teochef, has carried with him across the globe from his home province of Veneto. There, cooking is an earthy, meat oriented, but surprisingly sophisticated experience where these primal elements impress their signatures into the meal. Meat plays a big role in Teochef's cooking style, and he supports this love with claims of eating a decent steak at least three times a week.

Often our most memorable life lessons show us, by example, what we don't want. Having worked as a chef in Bassano and Cortina D'Ampezzo in Northern Italy it was in Paris, working for a year at one of the classical top restaurants, that Teochef received one of these lessons. There was almost no communication between him and his colleagues. Being Italian, his French co-workers just gave him rudimentary briefings in the morning, the barest of instructions, and left him there to his task unspoken to.

After that dark year, it was on to L.A. and Trattoria "La Botte", a star-spangled restaurant in Malibu, where he discovered the cross-cultural kitchen of California. This turned out to be the second major influence on Teochef's cooking style after his native Venetian cuisine. Ironically, during his relocation to the United States for this experience, the place where he went house hunting and then settled was Venice Beach.

Most everything one experiences at Cucina Teochef has the Raw vibe. The decorations, the menu, the tattoos, the man himself. From his soul music, heard coming from vinyl on a turntable in his office, the room Teochef refers to as "The Dungeon", to any given piece of restaurant decor or tool, they all have a story and a unique reason for being in the Cucina Teochef environs. It's hard to tell if the food is a reflection of the atmosphere or vice versa.

An unconventional cooking blend with pork and beef at the center. It is pork that Teochef favors. With an intimate knowledge of the pig, he can and will use all of the animal from nose to tail. He explains how his love of cooking has always been with him, "I've never wanted to do anything other than cooking, in fact, since I was little kid I've always been in the kitchen pulling on my mother's or grandmother's skirt asking them to have a spoonful of their homemade ragù that was still cooking on the stove. The art that is made in the kitchen has always been the reason why I wake up every day!"

EMBASSY OF THE SOUL AND PASSION OF BASSANO

MATTEO ZANUS

The kitchen where Teochef now makes that art is Cucina Teochef. After returning from LA in 2011 to Bassano, he held a few different head chef roles at various different restaurants in the historic city center. Bassano is Teochef's hometown, his city, his self-described heart and soul. It's where his family, in every sense of the word, lives. The biggest 'small town' in the area, it sits in the Veneto region and people come to visit from all over the world to taste the extensive gastronomic offerings found there. To sip on their world-renowned grappa, indulge in their history and to take pictures on that famous covered alpine bridge with the breathtaking foothills of the Dolomite mountains for a backdrop.

In 2014, he was offered a location near the Ponte Vecchio, but on the other side of the main city center, in an area called Angarano which at the time, was still quite virgin with regards to restaurants and any decent places to have a good meal. This is where he planted the Cucina Teochef flag, but it is by far not the end of the story.

To create the power of place needed, to foster the atmosphere Teochef felt necessary for his culinary creations, took passion and a clear vision. While some would be swayed by bills and bottom lines, Teochef stresses that his love for his calling comes first.

MATTEO ZANUS

MATTEO ZANUS

"I do my job exclusively for pleasure and when money is in the mix it becomes almost unnatural and venial. We must remember something, we live in Italy, the country of anti-business!"

Teochef fosters in his restaurant a respect for people, respect for the work and a respect for the mother earth who hosts us. The secret behind the freestyle and flow in that kitchen at Cucina Teochef is the humility of the people that work there, the dedication to this life that is often times, not so simple, and the sincerity in the dishes they propose.

This sincerity, trust and mutual respect can even be seen in Teochef's suppliers, or "extended family". The Fontana Meatbrothers, slaughterhouse and butchery are the one-stop, family-run business that supplies Cucina Teochef with all of their meat. The brothers provide a quality and attention to detail only possible because their slaughtering process is in-house, not often seen in a butcher these days. Teochef works with them closely to ensure the absolute highest quality meats and cuts for his creations.

MATTEO ZANUS

THE THING:

There are a few things, objects of necessity, or tools of the trade, that are integral to a chef's life and livelihood. The tools, shoes, and headgear that Teochef swears by, "Kai Shun kitchen knives, because Japanese steel handles well and is also extremely durable, Red Wings for shoes, comfort and the quality of the manufacturing is next to none, and a Stetson with regards to the headwear, no one is better than them and besides, they are the only ones who make a pigskin hat."

As we left, he imparted a few words of advice for new chefs, "Travel and work abroad, far away from your own country to learn about other types of food and traditions. Read and study."

For "non-Bassanoians" be aware that Venetian cuisine, much like Teochef, is not your average Italian fare and has very little to do with tomatoes, basil and mozzarella; even the pasta does not come typically with a "red sauce". For an original gastronomic goal, you must experience the foods of Bassano.

BACK TO NATURE

RAMON GRENDENE ON FREEDOM AND RESPONSIBILITY

Ramon Grendene studied design in Weimar, but soon after graduation, moved back to Egg, near Zurich, Switzerland. He grew up in a small village where his parents, an artist and a designer, bought an old farm which they lovingly restored. The living room and dining room are arranged around a large fireplace surrounded by beautiful objects, art, plants and custom dining tables and benches. The real treat, however, is the sweeping view from the side of the house. Across the rolling hills and a large pond, from time to time the smell of cows blows over this land and in the distance the sound of bells from flocks of sheep can be heard.

RAMON GRENDENE

In recent years, Ramon has dealt with two major projects: the construction and reconstruction of this house. Built and renovated by him, according to building biology principles with a view toward energetic self-sufficiency and associated overall system planning, according to the conceptual principles of permaculture. These design goals aim at constructing in harmony with permanently functioning, sustainable and natural cycles. Since coming back to this place, his life has changed and Ramon's new focus is making ecologically, economically and socially sustainable economies with the available resources.

SOIL IS THE FOUNDATION OF ALL LIFE

If the average garden could be called a song, Ramon's garden is a symphony. With everything perpetually in a state of experiment, based on the findings of permaculture. In the mid-1970s, the two Australians Bill Mollison and David Holmgren fostered concepts for the development of long-term productive agricultural systems as a sustainable alternative to the prevailing industrial agricultural methods.

In principle, they "rediscovered" the cycles of organic farming already known in Europe for themselves and their continent. They observed that industrial agriculture, with its preference for monocultures and massive use of pesticides, washes soil and water, reduces biodiversity, and erodes previously fertile soil. Today, such observations are confirmed worldwide and the conditions in the agricultural industry are increasingly criticized.

Globalization has changed many things. The seed of today is subject to the dictates of money. It's all about profit. This was recently demonstrated by the purchase of Monsanto by Bayer. Without seeds we have no harvest, no food. Because these small seeds carry life and nourish our lives, all ancient cultures held them in high esteem. Today, seeds are a business

RAMON RENDENE

RAMON GRENDENE

with 20 billion euros in sales per year worldwide. Three-quarters of this is made up of ten groups. The three largest of them share half of the market: Monsanto, DuPont-Pioneer and Syngenta. They fight for more and more market share with legal and breeding-related tricks.

THE ECO PRODUCTS IN BIO SUPERMARKETS TODAY ARE MAINLY HYBRIDS

With patents, corporations seek to secure ownership of their breeds and expensive royalties - not only for genetically modified seeds, but also for conventionally bred seeds. Hybrids are disposable seeds derived from inbred lines. With them, the seed companies get the best deal. Because hybrids provide good yields and consistent fruits. But these qualities will be lost in the next generation. The seeds of hybrids are not suitable for replication. Farmers and gardeners have to buy new seeds every year. Most do it - because the returns are high. Apart from lettuce and beans, many varieties of our vegetables, sunflowers and corn as well as some varieties of rapeseed and rye come from hybrids - also in organic farming. The big seed companies are also trying to transfer the hybrid technology to the breeding of wheat, barley and rice.

Ramon talked often to us of how important it is to deal with the topic of seeds, because the seed reflects life itself and the survival of the Earth dwellers depends on it. Naturally, seeds that have been optimally adapted to their respective location produce the strongest plants because the respective weather and soil conditions are stored over the years in the code of that seed.

Permaculture is the deliberate design and maintenance of agronomical and productive ecosystems that possess the diversity, stability and resilience of natural ecosystems. The philosophy behind permaculture is a view that works with and not against nature, an

RAMON GRENDENE

BOOGIE
ON

RAMON GRENDENE

attitude of continuous and deliberate observation, not of continuous and thoughtless action. Seeing systems in all their functions, rather than just demanding a kind of revenue from them, and allowing systems to demonstrate their own evolutions.

If one wanders through Ramon's garden, one immediately notices the biodiversity - the indispensable condition for the creation and preservation of ecosystems, as well as for a constant adaptability to evolutionary changes.

Horseshoe-shaped earth walls and wind-shielding hedges create microclimates that work wonders. Ramon knows a lot about planning according to the obviously stated differences seen. That it is important to observe and analyze a place according to its naturally prevalent features. Be they hot - cold, humid - dry, steep - flat, sunny or shady, with the aim of meeting the given conditions to include them in the design.

THE PRINCIPLE OF LESS-IS-MORE RESULTS IN MORE. ALWAYS.

A piece of what is here is the discovery of slowness. Long term instead of short term, less is more, against consumption in abundance. Optimizing instead of maximizing is a clear adage. So, ripen potatoes under hay, bean shrubs find support on corn stalks, pumpkin plants, green manure and mulch cover the ground. However, one thing here seems amazingly efficient and fast - composting. With great care, layer by layer of compost heaps are applied. Wood, leaves, green cuttings and compost-promoting plants such as comfrey, Jerusalem artichoke stalks and stinging nettles are chopped up, algae are added from the pond and biochar is added for nutrient fixation. Powdered with rock flour, the temperature of the cone rental will optimize within seven days and, once converted, will produce the most fertile soil possible. The most important basis for a completely organic production of excellent food.

When we visited Ramon, two young men were working on his farm. Over the week they slept in a teepee, and ate together under a tree by the pond. Ramon was teaching them all about, not only the individual components of a system, but in particular the relationships between them and their optimal use for building productive systems. So many of his thoughts go far beyond the creation of natural ecosystems into social components. More than ever, we are faced with the question of the right of self-determination for all people. Freedom and responsibility are among the great questions of our century. Ensuring everyone's right to freely use their livelihoods requires a balance between individual and community needs. This gives rise to an ethical demand for social justice. All people should have the same right of access to their livelihoods. On a small scale, much of this attitude and ethics can be found in Ramon's life. Our world needs more people like him.

RAMON GRENDENE

THE SCOTCH CONNECTION

BAR GABÁNYI & KÖNIGSQUELLE ON SINGLE MALTS

When asking for advice on Whisky it makes no sense to ask a young man. To find out about the myths of scotch making and tasting, one needs to seek counsel from men who are as matured as the spirits themselves. Hailing from the capital of beer, we went to Glasgow in search of direction on the 'water of life', only to return discovering that the very best we could find is right in our own town.

GABÁNYI & SCHREIBER

The two men we sought out were Peter Schreiber, co-owner of Königsquelle and Stefan Gabànyi, owner of Bar Gabànyi, both located in Munich.

Peter has been standing solid behind Königsquelle's bar for two decades, offering a smile and solidity to every patron. What counts for him is his team at Königsquelle, a Munich institution for good food and charming professional service and of course, good whisky from Scotland.

Stefan is not only a bar keeper but also the author of Schumann's Whisk(e)y Lexikon. He worked at Schumann's Bar in Munich for almost three decades then opened 'Bar Gabànyi' in 2012. This downstairs bar is like entering a different world, where the traveler is met with a perfectly mixed drink, good music and a hearty bite.

WHAT MONEY CANNOT BUY: A BARMAN'S ADVICE

At their bars, you get what you pay for, a quality drink. But what money cannot buy is the advice, "the consultancy" so to speak. This is where these two Makers excel, by creating a reliable anticipation of what to expect in your next glass, the dram, as the Scottish say. Today, more people drink single malt, but they drink it slower, a higher quality and typically in smaller quantities. Hence the ever-increasing value of a mature consultant as a drinking advisor.

GABÁNYI & SCHREIBER

EVERY GREAT WORK of ART
SHOULD BE SIGNED BY THE ARTIST.

Team Königsquelle
(v.l.t.r.) Christian, Peter, Sigi, Ranjit, Ralf
not present: Joschka and Franz

Single malt is not a new idea, but up to the 1950s all single malts were only used to blend. Only two brands, Glen Grant and Glenfiddich were actually producing singles for drinking. Glen Grant was at least available throughout the UK. Italy was the first country to give a push to the single malt market, and on the initiative of only one person, Glen Grant's distributor to Italy, Armando Giovinetti. He persuaded the distillery to sell him 50 cases and is quoted in Charles Maclean's Whiskypedia as saying, "My attitude was, if I don't sell it, I'll drink it." He sold it and what followed was a marked incline is sales as more distilleries jumped on the bandwagon.

ITALIANS WERE FIRST TO DISCOVER THE MERITS OF SINGLE MALT

According to Stefan and Peter alike, single malt is a drink to have while savoring the moment. One must calm down and drink it in humility. Humility for the age it took and the manufacturing process that went into it. It is a phenomenon that with almost identical ingredients that such ranges of different flavors are created by climate, age and the casks they were stored in to mature. Of course, Peter and Stefan will raise eyebrows here, as there is certainly more to it, but we will leave that for you, gentle reader, to find out.

The spirits in this book are recommendations. Some by the editors, for example the peated or smoky Talisker from the Isle of Skye (see page 280), an Irish Whiskey, Redbreat (see page 276) or the mild Bourbon Bulleit (page 248) or even single malts from Italy (Psenner's Rètico, page 274) or Germany (Stork Club Whiskey from the Spreewald, page 278) that offer refined qualitative alternatives to single malt scotch.

KÖNIGSQUELLE RECOMMENDATIONS:

1. GLENMORANGIE

The Original, 10 years

This one is a highland from picturesque Tain. 'The Original' is a very complex and balanced Malt with aromas of almond, honey, and citrus. An ideal companion for the early evening hours. Glenmorangie owns a spring known for its hard water and a guesthouse: "worth a stay!" (quote: Stefan Gabányi)

2. BRUICHLADDICH

By the legendary distiller Jim McEwan

McEwan worked for 53 years as a copper, a blender and distillery manager for various brands. Island malts usually come heavily peated but this one, from the Isle of Islay, has a slight cherry sweetness to it, subtly dominated by tobacco and wood.

3. MACALLAN

12 years

This 12-year-old from Speyside, like many Macallan whiskies, is matured in Oloroso Sherry casks. Giving it a fruity notion and oily appearance which provides a soft mouth-feel leaving slight vanilla and malt notes in the nose and on the palate.

4. AUCHENTOSHAN

Triple distilled, 12 years

One of the very few brands from the Lowlands of Scotland. This one is triple distilled and matured in Sherry casks, which results in a fruity, zesty, and sweet vanilla-toffee aroma. With 40% Alc. it weighs in rather mild.

PORT CHARLOTTE
HEAVILY PEATED
The MACALLA
HIGHLAND SINGLE MALT SCOTCH WHISKY
SINGLE MALT SCOTCH WHISKY
1
2
3
4

The Alambic Classique
1996
Tobermory
Alambic Classique
Collection
Rare & Old Selection
Highland Single Malt
Scotch Whisky
1977
Aged: 37 years
Glenturret
224 of 246
46,5% Vol. Product of Scotland 70cl
The Alambic Classique
Irish Diamonds
Small Batch
12 Years
70cl

Bar
Gabányi

BAR GABÀNYI RECOMMENDATIONS:

1. LINKWOOD

15 years

This old-school whisky from a negligently underrated distillery, is bottled by the renowned Gordon & MacPhail house. The taste is typically Speyside style with floral and lemon notes, tea and traces of white wine.

2. LONGROW

Peated

From the Springbank distillery in the western highlands, this is one of the few family farms that still cultivate traditional craftsmanship. Gabànyi's "Schuhmanns Whisk(e)y Lexikon" quotes the taste being "of wet sheep and the aggressiveness of a tiger".

3. ARDBEG

Uigedail

Located on Islay. Typically, Ardberg manufactures heavily peated whisky matured mainly in old bourbon casks. This heavyweight from the island convinces with strong smoky tones, slight sweetness and solid 54.2 percent by volume.

4. TOMATIN

18 years

Tomatin is Scotland's first distillery to extract much of its energy from biomass. Elegant and mature, this single malt highlander comes with a fine sherry aroma and is dangerously tasty, carrying a lightness defined by grassy and vanilla notes.

WHEN IN MUNICH

One final note, the distance between Königsquelle and Bar Gabànyi takes 30 minutes by foot. You can walk-off the meal and the first digestive scotch then pay the other a visit to sample more. A taxi takes 11 minutes. Rest assured every taxi driver in Munich knows their addresses!

Linkwood
15
Speyside
SINGLE MALT
SCOTCH WHISKY
GORDON & MACPHAIL
Longrow
Peated
Campbeltown
Single Malt
Scotch Whisky
Ardbeg
UIGEADAIL
1
2
3
4

CHAPTER ONE

PROVIDE & CULTIVATE

You can bring home the bacon ... or the fresh vegetables or the fruit, from the bounty of the forest, the sea or even a simple home garden. Makers, for generations, have brought us the keys that open the locks of dependency, allowing us to complete this circle of self-sufficiency.

SPAIN

BONAPROFIT

BRAND
Bon Aprofit is a food awareness project that aims to bring healthy food to people's lives through design

ESTABLISHED
2011

PRODUCTS
workshops

WEB
www.bonaprofit.com

THE STORY
Bon Aprofit started from a very simple question: "Is there a fun way to connect kids with fruits and vegetables?" The answer to that question ... edible tableware made from organic, local and seasonal fruits and vegetables.

The proposal is based on a series of workshops for children and parents, in which they make their own dishes. Quite literally, these kids and their parents make edible dishes as well as utensils, and cook a menu that includes the vegetables and the fruits used. Through the workshop children approach healthy food in a playful way, learning during the process the temporality, variety and physical properties of food.

For sustainability, Bon Aprofit adheres to an ecological and seasonal agriculture, fresh and local; which apart from producing high quality foods that benefit the health of its consumers, are great social and environmental bonuses.

BRAND
Family run farm with nearly 200 beehives and colonies

ESTABLISHED
2016

PRODUCTS
Honey

WEB
www.biohofrohrauer.at

THE STORY

A little south of Lake Neusiedl, in the middle of Burgenland, lies Lackendorf. Here the Rohrauer family cultivates a small apiary with a farm. Colorful orchids and green forests alternate around nearly 200 beehives and colonies. The history of the Biohof began with their great-grandfather, who, living in a time without a village shop, simply learned to do whatever he needed to supply himself and his family. He grew fruit, kept goats, and bred bees, whose honey he sold. Annemarie and Georg Rohrauer, a Berliner and a Burgenländer have run that farm since 2016 with the help of the whole family.

When bees fly home, they orient themselves to odors, the sun and also use the magnetic field of the earth. At the Biohof Rohrauer they make it even easier for their bees to find their way back. They've constructed and raised what they call flight boards above each colony. The flight boards of the individual bee colonies have different colors, so that the bees can see from far away where they need to bed down for the night.

Knowing each colony produced a unique honey, it became obvious to use the colors of the flight boards as their packaging labels.

BRISTOT

BRAND
Caffe Bristot is the leading brand of procaffé S.p.A, which has played a significant role in the development of Italian espresso

ESTABLISHED
1919

PRODUCTS
Coffee

WEB
www.caffebristot.com

THE STORY
In Belluno, nestled at the foot of the Dolomites, is Bristot. Founded in 1919 by Domenico Bristot, who was captivated by the alchemy of aromas achieved by means of skillful blends in coffees. That spirit led him to be one of the first coffee roasters to realize the great potential of developing the espresso machine, which has enjoyed growing success in public bars from the early 20th century onwards.

Inquiring, creative and never satisfied, Domenico continued his experiments and technical trials with new machines, working closely with inventors and manufacturers, in order to find the best high-quality coffee blends for these new extraction systems.
It was during this period that he formulated the blend that marked the culmination of his research, creating fruity notes and a chocolatey aftertaste in the cup. He named it Tiziano in honour of the versatile 16th-century painter from Belluno, Tiziano Vecellio.

The commitment to quality which is prerogative of small companies, and the capacity to buy advanced technology typical of larger companies, bring Bristot to uphold a high level of innovation and internationalization, giving it an ongoing competitive advantage.

For almost 100 years, creativity, "Italianness" and research have formed an intrinsic part of Bristot's DNA. What makes a perfect espresso? For Bristot, it's the people that make it. People from the mountains. Great workers who have learned how to live in harmony with the harshness of their region. Stubborn and tenacious, proud and passionate people. It is thanks to them that Bristot is a forward-thinking company today, renowned and appreciated for its fine quality in more than 65 countries worldwide.

BRONKO

BRAND
Apparel and accessoires made in the south of Chile

ESTABLISHED
2015

PRODUCTS
Denim, aprons, accessories

WEB
www.bronko.cl

THE STORY

Born from the dream of two lifelong friends, Nicolas Castillo and Gert Doering, Bronko adds a new flavor to the word 'craft'. Growing up together in the south of Chile, these young founders bring a unique vision through their partnership to Bronko products and as one of the team is now based in Germany, this brand spans continents, adding to the diversity of its product offering.

Bronko stands for tradition, quality and handcrafted goods, offering products that are skillfully and well constructed. This ethos is born out of a shared love for traditional craftsmanship and quality. Combining contemporary design with time-honored techniques, and offering well-architected products that are built to last, this shop uses only locally sourced materials, valuing the authentic and focusing on sustainability.

BRAND
Carabica offers fairly traded coffee made from all over the world

ESTABLISHED
2015

PRODUCTS
Specialty coffee beans and coffee brewing equipment

WEB
www.carabica.de

THE STORY
While most people drink coffee functionally and on the go, supporters of specialty coffees hold sacred every step from growing and preparing coffee cherries, to processing the beans, to roasting and skillfully defining parameters for the perfect cup of coffee.

The Makers at carabica revel in a fine coffee culture. They strive for that perfect cup of coffee, that magical shot of espresso and those cappuccinos that taste better than any dessert. However, they know that adhering to the rules of coffee preparation is not everything: above all, drinking coffee should be fun.

Specialty coffees come in lower volume and each offer a unique flavor profile. The aromatics in high end coffees are even more complex and varied than in wine. The ideal of specialty coffee is transparency and sustainability - in terms of taste as well as production. In search of more fairness in trading coffee, carabica believes that specialty coffees paired with direct trade relationships will dramatically and positively influence coffee growers' standards of living, touching entire communities.

DIRECT COFFEE

BRAND
Direct & fair trade coffee from Limmu mountains in Ethiopia, slowly roasted in Switzerland

ESTABLISHED
2015

PRODUCTS
Coffee beans, biodegradable capsules & cold brew

WEB
www.directcoffee.net

THE STORY
In November 2015 Marie and Michael Tuil came up with the idea of a social start-up that combines specialty coffee trade with positive change. Less than two months later they were already in Ethiopia, the home of forest coffee, meeting farmers, finding local partners for their social projects and selecting the first tons of coffee to import.

More than 10,000 varieties known as Ethiopian Heritage still grow wild in the Ethiopian mountain forests. Since the forest is so important for the variety of flavors and also for the price of coffee, the farmers protect the natural biodiversity. Simply put, forest coffee tastes better and protects nature! And because it grows exactly where the plant feels most comfortable, it does not need chemical fertilizers, pesticides or additional irrigation.

When it comes to buying the coffee, Direct Coffee's guiding principle is direct trade with smallholder farmers. The price they pay is not fixed in New York on the stock exchange, but in a personal conversation with the farmers on the ground. This brand wants to know the people who grow its coffee and make sure that they get their fair share. Direct trade also means no middlemen: this margin is instead invested in social projects benefiting the producing communities.

BRAND
An urban outdoor brand from bavaria for quality goods to venture out in

ESTABLISHED
2011

PRODUCTS
Urban outdoor tools and accessories

WEB
www.jeckybeng.com

THE STORY
Jeckybeng is part of an outdoor movement, inspired urbanites who see the world from a different perspective and are not afraid to question their everyday life. They continue to ask the fundamental question about why our modern society always seems to have to compete with nature instead of striving to achieve harmony.

With their interviews, stories, pictures and products, Jeckybeng are addressing all of us who are searching for deceleration, sustainability and balance – those who want to rediscover nature – to seamlessly step into the experience, taking only photos and leaving only footprints.

Products like their Universal Bag, offer the urban adventurer a perfect place to stash their snacks, breads or even smaller camping utilities. Featuring a unique roll top closing system, to keep nature out, when necessary.

Every day, the Makers at Jeckybeng are imagining and engineering new products that meet the practical requirements of wilderness and the esthetical preferences of a design-savvy generation. The products that they develop according to these wishes are designed to be a reliable companion in the city, traveling or simply outdoors.

JECKYBENG

BRAND
Outdoor brand built on the heritage of canoe-country ideals

ESTABLISHED
2001

PRODUCTS
Bags, luggage, accessories

WEB
www.frostriver.com

THE STORY

Frost River builds rugged, timeless packs and bags for a lifetime of use, with the highest quality waxed canvas, premium leathers, and solid brass hardware. All of the shop's products are made by hand in the USA and backed by a lifetime guarantee.

Every pack and every bag is made by Frost River in the manner most effective at delivering superior performance – The Old Way. The loggers and trappers, traders and adventurers who called the deep North Woods home – relied on their packs to perform in unrelenting environs. Frost River is built on that heritage. Their products are also reliable companions for hunting and fishing.

Today the ideas of living simply, off the land with what nature and your own skills can provide you, a true 'Life in the Wild' can take on an almost mythological sheen in the mind of the city nomad, but there are places left, and people that know how to get there, and what to do when you arrive. So it happened that Canadian cameraman Trustin Timber took a few friends along with Frost River products on a journey into the open vistas of unspoiled nature, bubbling mountain streams and campfires – he and his comrades lived the dream of getting away from it all: Catching fish at dawn and preparing it later on if skill and luck provided the catch of the day which is than cooked on an open fire in a simple manner. There is a film on Youtube "A Woodland Caribou Story" documenting Trustin's adventure worth seeing.

GRÄNSFORS BRUK

BRAND
The swedish forge sells its handcrafted axes to over 30 countries all over the world

ESTABLISHED
1902

PRODUCTS
Axes

WEB
www.gransforsbruk.com

THE STORY

For over 100 years, Gränsfors Bruk has developed an expertise and business around handcrafted axes with a focus on quality and environmental responsibility. When it comes to the peerless look and feel of their axes, much depends on the experience and skill of the individual craftsman. Therefore, each ax is marked with the smith's initials, a literal 'Maker's mark'.

One of the core goals of Gränsfors Bruk is to make unparalleled products that last a very long time. To prove that they take this promise very seriously, they sell all of their axes with a 20 Year Guarantee. All of their production process and materials are chosen with an environmentally friendly mindset, benefiting the employees and customers alike.

Located in the Swedish village of Gränsfors in northern Hälsingland, this forge sits at the heart of the village, and its rhythmic pounding can be heard, ringing through the air as you approach.

The primary forge is open to visitors, who can learn how axes are forged, sharpened, tempered, and fitted with handles. Apart from the forge, there is also a museum, where classes are given about the traditions of forging and log building.

HANDMADE FISHING FLOATS

BRAND
Made by one of the last professional float makers in England

ESTABLISHED
2014

PRODUCTS
Fishing floats

WEB
www.handmadefishingfloats.co.uk

THE STORY

Ian Lewis is one of only two remaining full time professional fishing float makers in the UK. In the warmth of the African continent, he discovered his passion for angling. After he returned to England, a hobby he had first tried as a boy, caught a fresh current. With the advent of the internet along with some pre-1960s fishing books, he started learning everything he could about old fishing floats and how to make them. Soon he was adapting patterns and making his own designs - often in collaboration with customers to fit particular fishing scenarios.Taking his inspiration from natural materials he works with, he's made a wide variety of fishing floats from patterns that would have been used by Izaak Walton in 1650 to more modern cutting-edge designs brought about through collaboration. Using 99% natural materials, including cork, balsa, bird quills, reed and even quail eggs, Ian has found these natural materials look and feel beautiful but beyond that, most are far more durable than modern plastic materials and should they break or get lost, are more biodegradable.

All floats are entirely handmade, from the cutting, shaping, gluing, painting and whipping. Several coats of custom lacquer are all applied by hand to give the floats a superior finish that is very durable. These heirloom pieces are designed so that they can be handed down to the next generation.

9mm
Balsa

HAFTSACHE

BRAND
Handcrafted items with a suprising story: these products are made by inmates of Bavarian penetentiaries

ESTABLISHED
2017

PRODUCTS
Cooking utensils, housewares, garden product, accessories

WEB
www.haftsache.de

THE STORY
The products of Haftsache have a special origin. They are produced by prisoners in Bavarian prisons. Made with careful handwork they are characterized by the very best materials and first-class quality - genuine craftsmanship.

Prisoners are part of society, though they spend time behind bars, they will eventually be released. These prisoners are preparing for the work of successfully entering society again while still in prison. Whoever buys a Haftsache, makes an important contribution to the successful reintegration of prisoners.

Deeply rooted in traditional handicraft, individual items of everyday use are created in the workshops of Bavarian detention centers. The products of Haftsache are very important due to the fact that the prisoners not only produce, but also become creative themselves and actively shape the goods. The handmade articles are not bulk goods which are produced on the production line. Each product is manufactured from the initial designs to completion. Thus, in every Haftsache there is not only genuine craftsmanship, but also a piece of the personal history of the prisoner who has worked on it.

BRAND
The last grain-mill in the greater Munich area still working today

ESTABLISHED
1703

PRODUCTS
Flour, grains, bread

WEB
www.hb-kunstmuehle.de

THE STORY

Everybody knows the Hofbräuhaus and that it's located in Munich. But, before it was re-built in the 1870s, it had its own mill attached to it.

Flour was the main product after 1870 when ground malt was no longer in demand. When the place was bought by a miller name Jakob Blum, it became a family-run business that is still in the hands of the Blum family today.

Originally, underground streams and rivers were used to generate power to run the wheels of the mill, and these waterways still exist today. But now, the mill is run by electricity and mainly produces flour. The little shop adjacent to the mill is like a time capsule, a place that feels very old except for the credit card reader.

There are guided tours every Friday that show the original way that bread was made starting from the grinding of wheat and other grains.

HARIO

BRAND
Outdoor focused, curated list of heirloom products

ESTABLISHED
1921

PRODUCTS
Coffee equipment, kitchenware, heat-proof glass products

WEB
www.hario.jp
www.heldbergs.com

THE STORY
The name HARIO translates to the King of Glass. For decades, this company has continued to deliver the beauty of color and new richness through high quality and heatproof glass. The foundation of Hario' s business strategy is 真善美楽 (shinzenbiraku). To pursue the truth of nature and well-being of people and fulfil needs which in turn become profit. To attain and make excellent use of high quality materials, mastered techniques and skilled personnel. To provide service and fine crafted products of value to customers. To deliver pleasure to society and provide happiness and wealth to businesses and employees involved.

The company has been dedicated to the design, production, and sale of heatproof glass since it was founded in 1921 and is still the only heatproof glass manufacturer in Japan to have a factory. At the time of founding, glass products for science laboratories and medical uses were produced and sold. After World War II, production of household items began starting with the coffee syphon. Since then, the company has expanded into the production of items such as their distinctive, gravity fed coffee filters, drip coffee pots and other coffee equipment, equipment for tea, Japanese teas and herb teas.

What's the story?

NETHERTON

BRAND
From the birthplace of the industrial revolution, this small foundry produces pots and pans that last a lifetime

ESTABLISHED
2011

PRODUCTS
Iron cookware

WEB
www.netherton-foundry.co.uk
www.heldbergs.com

THE STORY
It all began in the heat of smoky, smelting furnaces, with Abraham Darby I as he replaced traditional charcoal with coke. The truly timeless quality of his iron products was reaffirmed by his grandson Abraham Darby III, in 1779 as he surprised everyone with the first arched bridge built from self-made cast iron pieces. It still crosses the Severn River today. That same innovative Shropshire spirit also inspires the modern family company Netherton Foundry to make their kitchen treasures out of robust and extremely lightweight "spun iron". All of their pots and pans are classics, with attractive iron bodies, full of character, slim handles and handy lid knobs.

But this shop doesn't stop at superior product design, they are resolutely focused on sustainability. These days fields and forests have returned to Shropshire, and Netherton is convinced that it should stay that way.

They produce robust pots, pans and baking tins that are made to last, quite literally, forever. With a Netherton product the buyer purchases a companion for life, because there's no reason that such an outstanding pot will ever break. The secret of their special durability is combined with elegant functionality, and supported by cooperation between regional suppliers and the experienced craftsmen, who still hand-make their treasures.

HELDBERGS

BRAND
Carefully selected products for men that value quality and haven't lost their sense of adventure

ESTABLISHED
2015

PRODUCTS
Male focused, curated list of heirloom products

WEB
www.heldbergs.com

THE STORY
Everything seems possible in the imagination of a child: climbing the tallest tree, camping in the wilderness, catching fish and grilling them on the open fire – experiencing real adventures. But the things that little boys so vividly imagined sometimes seem far away in adult life. The Heldbergs brand was born to rouse that visionary of the past, and to signal that their old dreams don't need to be over. With selected products and distinctive one-offs, this curated list is aimed at men who are looking for something special and for whom "value" is not a foreign word.

Heldbergs speaks to those who aren't satisfied with mediocrity and mainstream. Their founder Volker Habermaass understands the needs for genuine personalities and passionate world travelers, as he himself is one. With Heldbergs he wanted to make men's dreams come true, make the things they need available, and delight their inner child. This knowledge is the basis of the brand, as the products offered are personally selected from around the world or designed in-house, and produced by exclusive manufacturers and traditional producers.

Men who are looking for something exclusive, special and unusual are in the right place. For a high-quality, sustainable world, for items that can be the great treasures of life. The adult world is already serious enough, and there are enough responsibilities with work and family. Inside every man there is a child who wants to play. Fun, after all, has nothing to do with age, but rather is a question of attitude, as well as the people and things that you surround yourself with.

HELDBERGS
What's the story?

HUNTER

BRAND
Iconic British rubber boot maker

ESTABLISHED
1856

PRODUCTS
Rubber boots

WEB
www.hunterboots.com

THE STORY

Hunter is a British heritage brand with a rich history of innovation and is renowned for its iconic Original boot. Established in 1856, the brand is worn by those who lead, forging designs to shield pioneers from the elements and the bleakest of landscapes.

The year 1956 marked the creation of the Original Green Wellington, the style that later became known as the Original boot, the iconic design at the heart of the Hunter brand. Today the boots are still made on the original patterns and are handcrafted from 28 parts.

Lauded for the quality of its footwear, Hunter was awarded a Royal Warrant by Appointment to HRH Duke of Edinburgh in 1977 and later to HM The Queen in 1986, an accolade of the highest order.

From pioneers and explorers to celebrities and icons, Hunter boots have soared in popularity. They were worn by Lady Diana Spencer in her engagement photos in 1981 and later by Kate Moss at Glastonbury. Today the boots are instantly recognisable on the festival fields of Britain and across the globe.

HUNTER

HUNTER
HUNTER

MOORMANN/CUSTOM-BUS

BRAND
A collaboration of two brands: Designer and manufacturer Nils Holger Moormann X bespoke engineering by Custom Bus

ESTABLISHED
1982

PRODUCTS
Nils Holger Moormann:
Furniture and custom interiors
Custom-Bus:
Custom Volkswagen vans

WEB
www.moormann.de
www.custom-bus.de

THE STORY
In Aschau, it seems a new wind is blowing. It has a mind of its own, but brings a breath of fresh air into the world of design. A pirate's flag, foisted under the Bavarian blue and white skies. Nils Holger Moormann positions himself somewhere between the conventions of conventionality and makes himself comfortable amidst his pieces of furniture and his custom interiors. From his cutting-edge living spaces to a luxuriously minimalistic tailor-made van created in co-operation with Custom-Bus, these living environments are not unlike the man himself: ascetic, well thought through, and with a nod and a wink to posterity that could be easily missed.

As an autodidact and a free spirit, Nils is the epitome of 'Neues Deutsches Design' and from 1982 onwards has developed with a number of mostly young, unknown designers as well as with his own creative department, a unique vision of furniture which exhibits a reduced formal language and precise solutions to detail.

The key thoughts here are simplicity, intelligence and innovation. They color the company's entire philosophy. At times this makes life uncomfortable, creates an edginess, but it speaks volumes for the timelessness and true Zeitgeist of the pieces themselves. And this is why almost all Nis Holger Moormann products are built by local artisans and sent on their way directly from the company headquarters.

LICHT AMBIENTE
an
aus
ANZEIGE WASSER/STROM
an
KÜHLSCHRANK
an

MORITZ WENZ

BRAND
A blend of modern and heritage designs curated and from its own workshop

ESTABLISHED
2001

PRODUCTS
Men's accessories, spirits, cooking utensils

WEB
www.moritz-wenz.de

THE STORY
This is a home to products and projects that have a history and tell stories, objects with high standards in terms of both their functionality and aesthetics, beautiful things, practical things and things of lasting value. The portfolio on offer includes fine spirits by small distilleries, men's accessories, sturdy rucksacks, cozy blankets, stationary and even sausages. The Makers at Moritz Wenz studios position themselves on the fine line between the quality of the past and the possibilities of the future.

This studio serves as an artisan platform, an interface between traditional manufacturers and modern entrepreneurship. The philosophy here is an honest focus on the craft production, as well as cooperation with small companies and family businesses. This is how the Makers at Moritz Wenz studio believe that qualitative products with history, soul and future must develop.

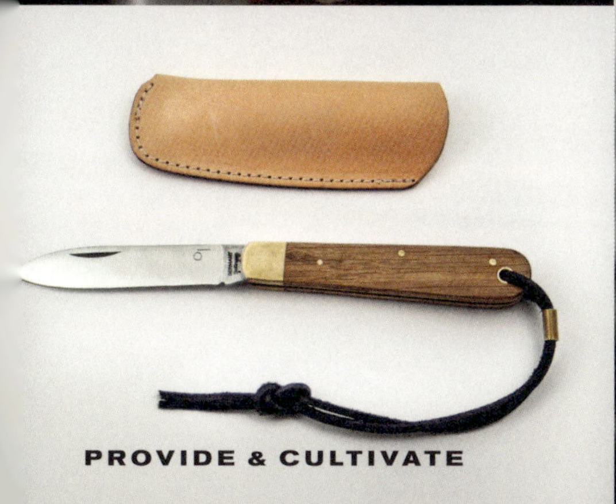

PASSION FRANCE

BRAND
Exclusive, handmade knives for pocket, table and kitchen

ESTABLISHED
2008

PRODUCTS
Knives and accessories

WEB
www.passionfrance.de

THE STORY
Once carried in everyone's pockets, blades and knives were the first tools in the history of mankind. Indispensable at work, in the fields, at home, and while traveling. Even today, knives still play the leading role in the kitchen and at the table and of course in industrial applications, as essential cutting tools in production.

It's unfortunate however that the smartphone, car keys and other every-day-carry objects of modern life have gradually displaced knives from their customary place. The Makers at PassionFrance are helping people to rediscover their appreciation for a knife in the pocket, a "compagnon fidèle", a reliable companion available in any time of need. This is especially true wherever people come together, wherever wine and bread is shared with friends.

PassionFrance products are 100% handmade in France, according to German quality standards. Every single knife is unique, each has an individual beauty and all connect their distinctive character with the "savoir vivre", the ease of the French way of living. And over time the knife will take on the character of the individual owner slowly changing to show how the knife is used, cared for and loved.

PassionFrance
BASQUE-YATAGAN
12 C 27
R.B.
PassionFrance
VENDETTA
12 C 27
DESIGN R.B.
LAGUIOLE
PassionFrance
DESIGN R.B.
DESIGN R.B.

RAWSON & PERRIN

BRAND
Rods, floats and reels handbuilt by passionate fishermen

ESTABLISHED
2009

PRODUCTS
Fly rods, split cane coarse rods, handmade reels, restorations, accessories, floats

WEB
www.rawsonandperrin.co.uk

THE STORY
When two fishing enthusiasts, who like making things, discovered their mutual passion, Rawson & Perrin was formed. Both men had been fishing from a young age due to their fathers' own love for the sport.

After discovering the joys of using centerpins and old cane rods together, there was no going back. Although carbon and fiberglass rods weigh less, the action and aesthetics of bamboo just amplified the joys of fishing for them.

Even though they had both restored some of their favorite rods in the past, it wasn't until Rawson began fly fishing that he seriously thought about building his own rods. Slowly, he began to gather the tools and equipment to begin building.

While his counterpart was dreaming of rod building, Perrin started to build and perfect his own floats. Rawson & Perrin kept their craft to themselves until 2009, when they decided to take their hobby more seriously.

The pair make everything by hand, using old and new techniques, which make for a stunning spectacle of piscatorial perfection.

RAWSON & PERRIN
MADE IN
ENGLAND
RAWSON & PERRIN
MADE IN
ENGLAND

SNEEBOER

BRAND
Hand-forged garden tools made to last a lifetime

ESTABLISHED
1913

PRODUCTS
Garden tools

WEB
www.sneeboer.com

THE STORY
Third generation Sneeboer family members Jaap Sneeboer and Wilma Peelen manage the company today, the 4th generation waiting in the wings, with the same philosophy their granddad started in 1913 – "if you produce a tool, make sure it's a good one"

The team, mainly composed of family members, are internationally renowned for the very highest quality, stainless-steel, hand-forged garden tools enjoyed by discerning gardeners.

The hallmark of Sneeboer is the synergy between new and old. Sneeboer artisans constantly seek new challenges, striving to develop innovative new tools, to improve on those tools that have proven their worth down the centuries and to resurrect tools which have been forgotten.

Skilled master craftsmen use traditional hand-forging techniques to transform new ideas into reality. New markets and modern methods are continually investigated and evaluated.

The next generation is now preparing to take over the company following the words of their beloved founder Arie Sneeboer: "I know how to improve." Another famous quote attributed to him is: "One who once has held a Sneeboer tool in his hands never wants anything else."

Located in the northwest of the Netherlands, Sneeboer manufacturing is famous for the highest quality Dutch garden tools available. These tools are enjoyed by avid gardeners in more than 20 countries around the world.

Old school hand forging is still the heart of every Sneeboer tool, but the challenge to improve and develop innovative products is never ending. Each tool is individually hand shaped, sharpened and polished with a superb eye for detail. With an eye on our future, Sneeboer is using "Green" practices. For example, they are fitting all their tools with FSC Certified hardwood handles.

SCHMIEDEWERK ASBECK

BRAND
Old blacksmith craftsmanship made by young hands

ESTABLISHED
2012

PRODUCTS
Knives

WEB
www.schmiedewerk-asbeck.de

THE STORY
Traditionally made, high quality, handmade knives, designed to each customer's specifications is what Schmiedewerk Asbeck stands for.

Growing up on a rural horse farm, Benjamin Asbeck started making art at an early age, learning how to craft knives from his parents. Now, living in the beautiful Chiemsee region of the Bavarian Alps, Benjamin finds his inspiration in the elemental nature surrounding him while crafting his knives. Pairing art and knife-making while using only recycled materials for forging the blades, each knife is uniquely made for each customer. The knife is as unique and special as its owner, who could be a hunter, fisherman, chef, or traditional outdoorsman.

Love of the craft, nature inspired, and ecologically-friendly, these knives are made to be special, as individual as each owner.

BRAND
Leather goods and decoration pieces made in Munich

ESTABLISHED
2012

PRODUCTS
Aprons, bags, leather goods

WEB
www.studiomunique.com

THE STORY
For the love of things that are beautiful, functional and handcrafted, Studio Munique was born. A unique design space founded by Makers, Stefan Geisberger and Susanne Ulhorn. Their primary goal is to make leather goods which last a long time. And their recipe for achieving this is choosing materials which feel good to the touch and will never wear out, like waxed hemp, glazed cotton, selvedge denim and recycled or upcycled materials. In order to maintain quality and the unique craftsmanship character of the brand, each of their products is finished in the Studio.

Words like quality, and craftsmanship cannot be over used in relation to Studio Munique. The Makers that create these essential accessories for fashion, home and living, do so with a passion and skill that shows through from every extra stitch in their denim and leather aprons, to the patinas and natural patterns on the brass and leather that make up many of their other products.

STUDIO MUNIQUE

STANLEY

BRAND
Stanley is providing warmth whenever you need it and comes with a lifetime warranty

ESTABLISHED
1913

PRODUCTS
Vacuum bottles, food jars, mugs, coolers, jugs

WEB
www.stanley-pmi.com

THE STORY
William Stanley Jr. forever changed the way we transport and consume both hot and cold beverages. In 1913, Stanley fused vacuum insulation and the strength of steel into one portable bottle, inventing the all-steel unbreakable vacuum bottle we know and love today.

In the century since, Stanley's vacuum bottle evolved from concept to icon and has become an essential part of workdays, road trips and outdoor adventures. From thermos to cup to even flask, these incredibly durable, hard-core metal containers have become a boon companion in our everyday lives.

Stanley has powered daily coffee breaks and just as many meals in their food and beverage gear. With each passing year they continue to push themselves to bring timeless yet modern solutions to today's outdoor adventures. For these Makers, it's about ramping up your time outdoors with gear you can count on from dusk till dawn. With coffee to kick start the day and spirits to cap off the evening.

STANLEY
MASTER SERIES

VESCOVI

BRAND
Vescovi uses up to seven different types of Arabica Gourmet coffees

ESTABLISHED
1927

PRODUCTS
Coffee

WEB
www.caffevescovi.it

THE STORY
If coffee is an art, the Makers at Vescovi are the artists of your everyday life. For more than 80 years, the Premiata Torrefazione Vescovi has trusted in its dedicated and passionate master roasters to obtain and roast only the perfect coffee, creating the unique Grani d'Oro blend. Grani d'Oro is a dual blend coffee, based on an original recipe of the founder, the Commendatore Gino Vescovi.

Imported directly from the best areas of world production, their blends are the result of traditional recipes and careful selections of the very finest Arabica quality beans. This extraordinary blend is made up of seven different types of Arabica Gourmet coffees.

By cultivating long term relationships with the farmers that produce these beans, Vescovi sets a consistency of flavor and texture, giving their product a blend of constant characteristics one can depend on and enabling them to guarantee quality.

The village is a place where traditional products are true treasures and through expert hands, they become the benchmark everyone strives for. It's the same vision that has been part of the Premiata Torrefazione Vescovi since 1927.

Vescovi

WALDKAUZ

BRAND
Originally made only for personal use, the hunting gear and outdoor bags have since found many fans and customers

ESTABLISHED
2009

PRODUCTS
Hunting gear and accessories, leather bags, clothing, gun sleeves

WEB
www.waldkauz.net

THE STORY
During a nightly hunt, Waldkauz founder Markus Holthausen came up with an idea for an extraordinary bag to organize his equipment. The very next day, he made his idea a reality using an old loden leather blanket and sewing machine.

This self-made prototype was no beauty, but it fulfilled its purpose perfectly and aroused the curiosity of other hunters. All this interest prompted him to work out a more professional pattern with a real seamstress. Soon after, he founded his own brand and started selling to interested hunters.

Inspired by initial feedback, he and his wife Gudela and their growing staff of seamstresses developed new products for hunters under the label Waldkauz using the logo of the tawny owl.

Later on, he expanded his business into the new label Steinkauz, specializing in outdoor and leisure equipment and clothing, made of natural fabrics.

With an undying spirit for innovation, fine sense of design and a sophisticated quality, the small manufacturer develops timeless bags, accessories and clothing from loden, felt and robust nubuck leather. With the leather scars and irregularities, each item is a unique piece which ages with dignity in the hands of its owner.

WALDKAUZ

&SONS

BRAND
Unique workwear clothing designed by and for craftspeople

ESTABLISHED
2016

PRODUCTS
Clothing, workwear

WEB
www.andsons.co.uk

THE STORY

The &SONS project is the creation of a range of unique workwear clothing designed by and for craftspeople, artisan makers and creative pioneers in their chosen fields of endeavor. This clothing is the brainchild of a group of like-minded craftspeople who make their living from taking and shaping superb visual imagery.

When it comes to creating a brand of clothing that reflects traditional values, employs time-honored skills and uses only the finest natural materials, the Makers at &SONS knew how to take on that challenge. They paid such close attention to sourcing from only the best providers of tailored clothing with care in every stitch and love in every seam.

The result is a range of distinctive clothing items that draw on the past to create new and meaningful garments and accessories that you will be proud to possess and wear. This is clothing for creative pioneers: individuals who stretch themselves, their boundaries and their horizons. These are clothes for people who are cut from a different cloth.

SHADOWPLAY
Coffee.
Edit.
Save.
Repeat.

&SONS
&SONS

CUPMASTERS OF THE UNIVERSE

HORST A. FRIEDRICHS

MEETING BARISTAS AROUND THE WORLD

Our ever-advancing technology, coupled with an increased understanding of how sustainability and transparency in supply chains can produce specific taste experiences when it comes to this beautiful little bean, have driven end to end solutions for growing, roasting, preparing and finally enjoying coffee, not as a simple pick-me-up, but as a fully experiential event for the senses. A best-selling tale whose characters are often as unique as the rich brown liquid that crosses our lips. A drink that has become an art in its own right.

HORST A. FRIEDRICHS

HORST A. FRIEDRICHS

HORST A. FRIEDRICHS

HORST A. FRIEDRICHS

HORST A. FRIEDRICHS

HORST A. FRIEDRICHS

HORST A. FRIEDRICHS

CHAPTER TWO

CUT, COOK & BURN

It is said that Prometheus brought mankind fire, the fire of technology. Tools to prep and ways to heat sustenance, empowering us to outstrip the beasts of the wild and carve a place for ourselves on this planet. That fire and burning passion are reflected in these artifacts.

BEEFER

BRAND
Pioneer in high-end meat grilling at 800°C

ESTABLISHED
2011

PRODUCTS
Grills

WEB
www.beefer.com

THE STORY

In the summer of 2011, Frank Hecker was inspired. His visit to a world-famous New York steak house fired his passion hot enough to convince his friends Marc Kirwald and Frantz Konzen of the idea of developing a high-temperature grill for beef that precisely generates the unique taste he fell in love with. This grill could be used at home in the garden or on the domestic veranda.

Cooking beef with ultra-high 800 degrees upper heat has become one of the most spectacular trends. Not only in the high-end grilling scene, but throughout the gourmet kitchens of the world. The extremely high temperatures in the Beefer grill are able to finish the meat in a very short time with unique roasting aromas and an incomparable crust - while at the same time the interior of the meat remains untouched, juicy and tender. The result is an almost indescribable taste explosion.

Since the heat acts without contact from the top of special high-performance ceramic burners, fat and meat juices do not drip unintentionally into open flames, where they can be burnt and thereby produce harmful by-products. Instead, caught in a bowl, they form the precious raw material for sensational sauces.

The original Beefer, the world's first home, gas grill certified for 800 degrees celsius, is not limited to beef. Whether it's fish or seafood, vegetables or desserts, the Beefer conjures up a crisp texture with a unique variety of aromas in just a few seconds. In other words, the Beefer opens up completely new dimensions to culinary delights.

To this day, the Beefer, despite its great success, is still hand-made in stainless steel, with an absolutely puristic design and 100% "Made in Germany". In Bad Honnef, the romantic Rhine town at the foot of the famous "Dragon Rock". And in a workshop for disabled people in Bonn.

Beyond the "Ur-Beefer", launched in 2014 the inventors and founders of Beefer Grillgeräte have developed three new variants. The equally large "Beefer One Pro", which has a comfortable electric ignition. Also, two double-sized "XL" versions. The flagship is now the "Beefer XL Chef" with approval even for indoor use in professional kitchens.

B

MediaRange

BERKEL

BRAND
Slicing in mechanical perfection with "La macchina"

ESTABLISHED
1898

PRODUCTS
Slicing machines

WEB
www.theberkelworld.com
www.tba-hausgeraete.de

THE STORY
Innovation that guarantees simplicity and efficiency. A myth that came into existence back in 1898, thanks to Wilhelmus Adrianus Van Berkel, who used his vast experience to mechanically reproduce the movement of the hand that cuts with a knife. During the `50s, Berkel was not just a pioneering product but also a true status symbol.

This pursuit of perfection that translates into the highest attention to detail is only possible thanks to an entirely artisanal production process. The high-precision mechanical features enhance the quality and blend with the beauty of these devices, only heightening the extraordinary appeal of Berkel slicers.

These professional machines are designed to ensure maximum reliability and precision, using pioneering solutions and leading-edge technology. Each Berkel design is rich in details added specifically to obtain the highest level of safety and minimize waste.

The elegance of shapes, the harmony of movement, the brightness of blades, and the Berkel red, all combine to make a true design icon. Subject to imitation as one of the most renowned luxury brands, Berkel brings all the charm of history to your home. Architects and interior designers seek out these tools to bring a unique touch to warm, traditional kitchens and living rooms or create a vintage atmosphere in modern environments.

After over 100 years, Berkel is not just a trademark, it's a lifestyle, a passion.

Berkel

BIG GREEN EGG

BRAND
Based on the design of Japanese Kamados, the Big Green Egg combines ancient techniques with NASA materials

ESTABLISHED
1974

PRODUCTS
Ceramic grills

WEB
www.biggreenegg.eu

THE STORY
The Big Green Egg is based on a 3,000-year-old Asian clay oven - a traditional wood-fired oven that, even in its earliest days, created surprisingly pleasant results when it came to taste. This ancient design combines production processes and innovative materials with modern knowledge to create the ultimate cooking device and experience. The high-quality ceramic ensures very low fuel consumption while the lid is on. This is thanks in part to the perfect circulation of air that ensures the food cooks evenly and at just the right temperature. This kind of cooking balance helps the Big Green Egg to bring you exciting, delicious and juicy meals to the table, every time.

The beautiful, functional design of the EGG and the use of superior materials mean that the Big Green Egg is top tier. This cooking device is produced from a very exclusive and extremely high-quality ceramic technology that was advanced by NASA. No other cooker is as reliable, sustainable, weather-resistant and heat-insulated. Furthermore, the heat bounces back off the ceramic, creating an air flow that gives an exceptional taste to all food and dishes that you prepare in the EGG. The result is the ultimate taste experience.

As it is highly reliable, you can enjoy the Big Green Egg worry-free. The easily regulated temperature is very stable. Because of the high-quality heat-insulating ceramic, external temperatures do not affect the temperature inside the EGG. The two adjustable vents - the draft door and the dual function metal top - make it possible to accurately set and maintain the temperature to within a few degrees. The smaller the openings, the lower the temperature, and vice versa.

Big Green Egg has a temperature range of 70-350°C. This contributes to the fact that, sometimes in combination with certain accessories, the Big Green Egg can be used for all manner of cooking techniques, including grilling, baking, boiling, stewing, smoking and slow cooking. The aroma it bestows upon your dishes will surprise you and your guests.

Big Green Egg
Big Green Egg

BULTHAUP ACCESSOIRES

BRAND
Luxury kitchen accessoires with an emphasis on design

ESTABLISHED
1949

PRODUCTS
Kitchen accessories

WEB
www.bulthaup.com

THE STORY
bulthaup has a full range of additions to make any kitchen feel like your kitchen and these accessories are more than just cooking tools. We use them every day, from slicing bread and prepping vegetables, to measuring or storing flour and spices, even serving up our meals. Well made, quality accessories gain in personal value the longer we have them. They begin to retain a kind of fingerprint or impression of our personality. bulthaup's characteristic expectations of design, material, and workmanship make that connection effortless.

Wood remains the first choice in the kitchen for bulthaup, especially when preparing food with knives. The wooden board develops an individual patina with daily use. At the same time, ingredients within the maple wood have an anti-bacterial effect. Also chosen for many of the containers, wooden and stainless steel drawers are ideal accessories for people keen to use the storage space in their kitchen right down to the very last corner. They fit perfectly into bulthaup pull-outs and are especially ideal for storing.

bulthaup

bulthaup

CARL MERTENS

BRAND
Carl Mertens stands for well designed cutlery and premium interior accessories

ESTABLISHED
1919

PRODUCTS
Cooking and kitchen knives, cutlery, dining table and home accessories

WEB
www.carl-mertens.com

THE STORY
In 1919, Carl Mertens was founded as a commercial grinding shop for pocket knife blades. Traditional craftsmanship and an uncompromising demand for high quality combined with innovative designs are the key factors that distinguish the collections of Carl Mertens. With new investors in May of 2015, the freshly forged Carl Mertens International GmbH initiated extensive restructuring: the traditional company developed from being a classic tableware brand into a high-quality manufacturer for individual lifestyle around the fields of preparation, cooking and indulgence.

Going forward the main emphasis has been on products that represent an active and individual lifestyle. Life in the kitchen and all around the set table – inside as well as outside at a BBQ – are areas of life in which Carl Mertens products can be found. By combining elaborate design with traditional craftsmanship, professional manufacturing and a high level of convenience, the company transports decades of know-how in the processing of stainless steel. Carl Mertens products tell their story of manufacturing, material and a truly unique form.

Cooking and dining is a journey full of adventures and new discoveries. For almost 100 years, Carl Mertens products have been creating helpful tools for this journey.

METZ
METZ

BELGISCHER BROCKEN

BRAND
The hardest cutting stone for the sharpest knives

ESTABLISHED
1999

PRODUCTS
Grindstones, sharpening stones

WEB
www.belgischerbrocken.de

THE STORY
The Belgian Brocken is a grindstone like no other. Even the ancient Romans knew and appreciated the unique qualities and measurable results received when putting an edge to their weapon of choice by using this special rock.

To this day, the Belgian Brocken is regarded as the finest grinding stone in the world. Since 1625, it has been mined exclusively in a small region in the Ardennes. This ancient area is around 480 million years old and quite distinctive. These Belgian boulders consist of volcanic ash, dense slate and lots of fine granites. The complex interplay and layering of that material ensures the highest quality and precision when sharpening. The Maker's at Belgischer Brocken know this and know how to form, cut and combine their stones to help anyone get the edge they desire.

BRAND
Exclusive cutting boards for the modern kitchen

ESTABLISHED
2015

PRODUCTS
Mise-en-place cutting board

WEB
www.cuttworxs.de

THE STORY
Cuttworxs cutting boards are the ultimate accessory for the modern, sophisticated kitchen. The exclusive mise-en-place workstation is ideal for the kitchen, as well as for the BBQ.

The Cuttworxs brand was born in 2015, but Carsten Schneider's innovations go much further back in time. As a trained mechanic, Carsten was skilled in working metal, and because he is an avid BBQ fan, he soon began building his own custom designed grills. He came into fame via Facebook where hard core grill fans in Germany and beyond discovered his specialty grills. These fans revealed dissatisfaction for the preparation boards and stations that were available, and this information fueled Carsten's development of the Cuttworxs cutting board and brand.

Self-taught and with countless local and national chefs in his customer base, Carsten is still a humble man from Saarland, Germany. Although he has reason to be proud, since he's won the 2018 German Design Award for "Best Kitchen Product". He did raise an eyebrow at this and celebrated with friends, beer and BBQ.

DUALIT

BRAND
Iconic British household appliances

ESTABLISHED
1945

PRODUCTS
Toasters, kettles, food processors, waffle irons, juicers and grills

WEB
www.dualit.com

THE STORY
Max Gort-Barten was a German-born entrepreneur who found himself in England, just as the second World War broke out. He, of course, signed up for the British army. When the war was over, he bought a small factory in Camberwell and began bringing several of his ideas to fruition. These included the Dual-Light electric fire from which Dualit took its name, but it was a patented toaster that really caught people's attention.

Max designed and engineered a six-slice toaster with a built-in timer (the first of its kind) in 1952, which bears a striking resemblance to today's Vario toaster. When the toaster became the must-have kitchen appliance of the '80s, the company was swamped by an insatiable demand. Today Dualit produces a wide range of kitchen and home appliances. It is run by Max's son Leslie and is one of a handful of successful, independently-owned British manufacturers. Reliable, well-engineered products remain at the heart of the company and there is an instantly recognizable, nostalgically retro Dualit aesthetic which can be traced back to Max's first toaster designs in the 1950s.

These Classic Toasters have been hand built in the UK since the 1950s. Featuring fully replaceable or repairable parts, timeless styling and the unique, patented Proheat element. No frills, no gimmicks, no compromise, just great products!

Dualit

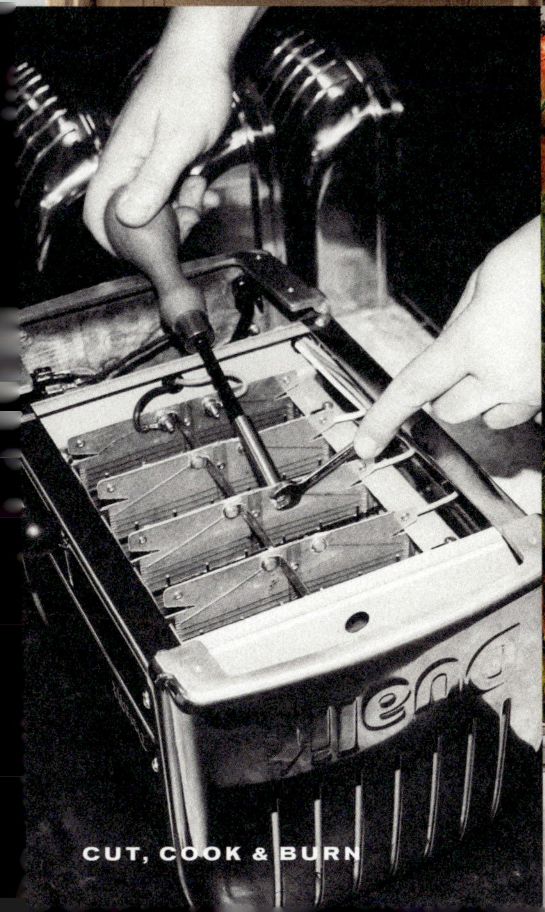

MAX
1.7L
1.5L
1.25L
1.0L
0.75L
0.5L
0.25L
Dualit

EPICUREAN

BRAND
Epicurean offers a full line of sustainable kitchen products for the home gourmet

ESTABLISHED
2004

PRODUCTS
Cutting surfaces, utensils

WEB
www.epicureanusa.com

THE STORY
Epicurean is a portrait of the real American Dream. The founders, a trio of young, scrappy, hard-working entrepreneurs from Minnesota, started a business designing and building advanced, composite skate parks for communities across the country. TrueRide would go on to build more than 400 skate parks.

With an ever-present future eye towards environmental issues and responsible recycling, the founders started taking a close look at the waste and off-cuts generated from the fabrication and building of the parks. They quickly realized that they could make new products from the remains and Epicurean was born. What started as a few cutting boards given to friends and family soon exploded into a business with a life of its own.

The beauty and brawn of Richlite, the material used in the boards, revolutionized the cutting board market. Richlite has the best qualities of both worlds with the warm look and feel of wood and the dishwasher-withstanding ability of plastic. Epicurean set a new standard for premium cutting surfaces in the home and in the professional kitchen. They continue to create state-of-the-art, market-leading, heirloom products proudly made in the U.S.A.

FEUERRING

BRAND
Simply shaped barbecue and fire bowl made in Switzerland

ESTABLISHED
2009

PRODUCTS
Grills

WEB
www.feuerring.ch

THE STORY
Since its launch in 2009, the Original Feuerring has opened up new dimensions in the barbecue market. Today, when looking at the number of copycat products in Switzerland, and indeed in the rest of Europe, one could almost speak of the need for a new "revolution". While the barbecue market tries to regroup after the seismic shift triggered with indirect grilling by Feuerring, the "first of its kind", the original itself is striking out powerfully on new paths with its inherent creativity, authenticity and the awareness of such values as quality, sustainability, mindfulness and health.

With the initial idea – in the form of a sketch from steel sculptor Andreas Reichlin, they teamed with Heiri Gisler in Eastern Switzerland, an expert in the area of sheet metal forming. Based on that simple drawing, he made the first wooden buck in beech wood. The 6 mm thick metal sheets were then pressed onto it cold, using a pressure roller. This takes a sure instinct and experience as well as a ferrous alloy which doesn't crack under tension or wrinkle when compressed, so that the bowls are perfectly-formed.

All the bowls are transported to Küssnacht am Rigi, to the company Isenschmid AG. Together with Andreas Reichlin, they make every effort to produce the perfect steel surface for the grill ring. Only the most attractive surfaces and the best sheet metals are selected and cut for the Feuerring and embossed with the Original stamp.

Using their patented technique, the grill ring and the bowls are welded inside and out. The welding and the angle ensure that even with major temperature fluctuations, the Feuerring will retain its shape, making it an heirloom.

Once your first fire is burning – you can prepare the food for grilling – depending on how fierce the fire is, you can begin barbecuing within 30 to 45 minutes. At this stage, your Feuerring is around 572 °F towards the centre and ca. 302 °F towards the outer edge. You can use these temperature zones to match the food on the grill: cook raclette cheese on the very outer edge, sear beef at the inner edge or slow-cook a whole fish on an additional wooden board... to suit your own taste, taking your time and at your leisure.

BRAND
Founded in Tel-Aviv's Florentine neighborhood, the knives by the eponymous brand are hand-crafted tools from Barcelona

ESTABLISHED
2012

PRODUCTS
Cutlery

WEB
www.florentinekitchenknives.com

THE STORY
Florentine Kitchen Knives is a design and manufacturing company dedicated to the production of high end cutlery by incorporating traditional knife-making methods with modern-day technology. The company, founded in 2012 by product designer Tomer Botner, was originally formed in the neighborhood of Florentine in south-west Tel-Aviv, where the company got its name. It has since moved to Barcelona, Spain and in early 2018 will be opening their first independent shop and Atelier in the heart of Barcelona where visitors will get to see the production process up close and even order their custom knives.

The company's signature handle design, which is based on an old leather handle stacking technique, makes their knives stand out and their use of high end materials and attention to details has made them the knife of choice by many chefs and cooks around the world. FKK's table knives also grace the dining tables of some of the world's finest restaurants where they produce different color and material patterns unique to each establishment.

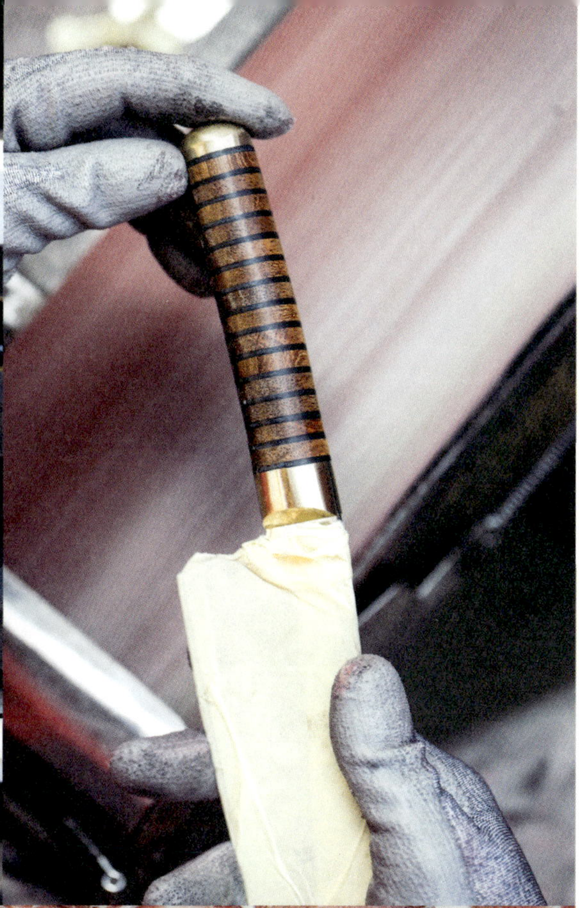

FRIEDR. DICK

BRAND
Quality knives and tools for professionals worldwide

ESTABLISHED
1778

PRODUCTS
Knives and accessories, steel sharpening machines, files and rasps, sausage makers

WEB
www.dick.de

THE STORY
The history of Friedr. Dick spans 235 years from its foundation in 1778 as a file cutting workshop in the southern German town of Esslingen. The company goal was to provide quality products at home in Germany and eventually globally. And they succeeded! The small workshop grew into an innovative industrial company, now based in Deizisau, which despite its global success has remained true to its roots. Friedr. Dick is the only manufacturer worldwide offering a complete range of knives, sharpening tools, grinding machines and accessories for butchers and chefs. And it's still a family-owned company after all these years.

Friedr. Dick products are created with sustainability in mind, using traditional manufacturing methods and state of the art production technology. This shop is well known to professional chefs around the world as a lead innovator in this field. Its international reputation as a manufacturer of quality utensils is based on consistent reliability and durability of the products.

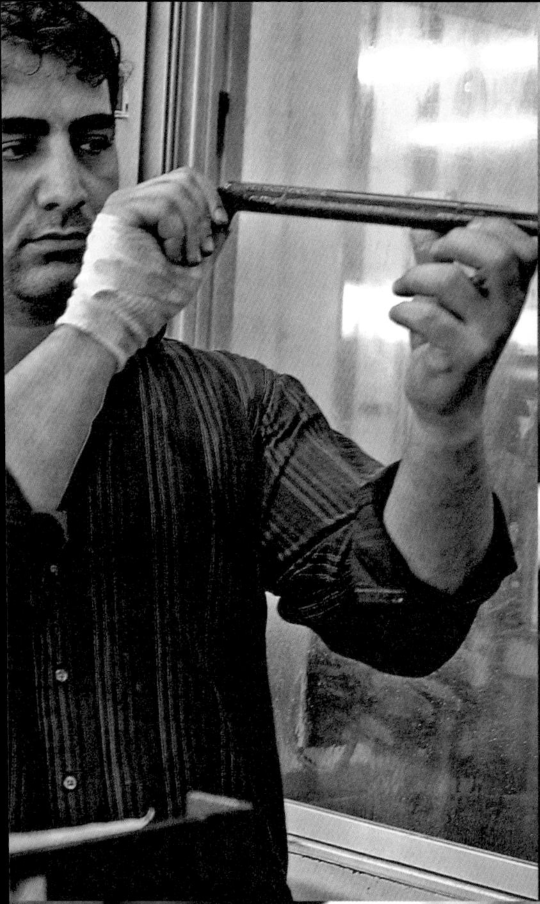

F.DICK
Made in Germany
Made in Germany

GÜDE

BRAND

Knives in small numbers and of the highest quality

ESTABLISHED

1910

PRODUCTS

Knives

WEB

www.guede-solingen.de

THE STORY

The cooking knife has been reinvented with the Solinger traditional blade-smith Güde and their flagship creation, 'The Knife'. This family owned business has, in its fourth generation, achieved success in innovation, in form and in function. The Chicago Athenaeum has also recognized this achievement and awarded 'The Knife' with the Good Design Award 2016/2017. Founded in Chicago in 1950, the Good Design™ Award is the oldest, most prestigious and globally recognized award for excellent design.

With the design of 'The Knife', Güde had the courage to take a whole new path in cutlery creation. Due to its exceptional shape, this knife becomes an almost natural extension of the hand. In this way, any chef has perfect control over a perfect tool.

Solingen is known for the highest quality blacksmithing craft made in Germany. Conventional knives consist of three parts: blade, crop and handle. 'The Knife' is a harmonious whole, whose blade extends seamlessly into the handle over the broad back. Such a breathtaking shape can only be achieved with the traditional drop forging technique. For a real lover of knives, Güde has a special edition of 'The Knife' which contains more than 300 layers of damascus steel.

HOFFMANN METALLGEFÄSSE

BRAND
Cast iron kitchenware that is in equal parts sculpture and cooking utensil

ESTABLISHED
1985

PRODUCTS
Cast iron cookware, tableware

WEB
www.hoffmann-metallgefaesse.de

THE STORY
Since 1985 in the studio of Berthold Hoffmann, individualistic objects of the cooking and table culture have been developed and manufactured in small-scale serial production. Mastering artisanal techniques of metal processing has enabled Mr. Hoffmann to apply his design ideas directly to sculptural objects of everyday life.

Pans, pots and roasters are developed in his studio and made according to his casting models in cooperation with regional iron foundries. The reworking or final finishing of castings in his workshop marks the individual character of each item. The surfaces of his iron pieces are burnt black with linseed oil. As a result, you obtain a unique product which will last a lifetime.

The silverware is manufactured by Mr. Hoffmann after working an idea out in sketches, paper models and then working samples in base metals. From idea to finished object, a variety of craft techniques are used. Just some of which are: forging, annealing, brazing, grinding and polishing. These heirloom products are made of solid sterling silver. The material thickness of the semi-finished utensil or tool is usually between 1-1.2 mm giving you a high-quality object for everyday use.

WEYERSBERG

BRAND
Cookware made of copper in combination with modern technology

ESTABLISHED
2009

PRODUCTS
Pots, pans, cookware

WEB
www.kupfermanufaktur.com

THE STORY

After more than 30 years of creating traditional copper cookware, in 2009 these artisans gathered new investors, and Kupfermanufaktur Weyersberg was born. At the headquarters in Baden Württemberg Germany, these specialists, with their great sense of quality, perfection and experience in copper processing, remain more active than ever.

In contrast to earlier times, when copper was tinned, these products have an innovative surface technology with a specially developed "Premium CeraSigno seal". This black ceramic coating provides an excellent non-stick effect and it has the perfect roasting characteristics. The visionaries at Kupfermanufaktur Weyersberg have also developed a thin ferromagnetic bottom coating, which combines the excellent temperature regulation of induction stoves with the perfect heat distribution of copper. So, every pot and pan offers superb functionality coupled with unique design and a great overall feel.

The physical properties of copper guarantee an even, fast heat distribution of the material as the conductivity of copper is 25 times higher than stainless steel, and 9 times higher than aluminum. This fast reaction is especially helpful when precise temperatures and cooking times have to be observed.

These products offer a bit of everyday luxury for those who want to feel, see and taste the quality while experiencing the beauty of optimal cooking tools.

MOCCAMASTER

BRAND
Reliable and sustainable hand-made coffee machines

ESTABLISHED
1964

PRODUCTS
Filter coffee machines

WEB
www.moccamaster.eu

THE STORY
This Maker's legacy of reliable and sustainable coffee brewers began with the release of the first Moccamaster designed by Dutch industry designer Mr. Smit. Styles and trends have changed over time, but Moccamaster brewers continue to set the benchmark for perfectly brewed filter coffee. Thus they have been awarded best filter coffee brewers in the world many times. The factory, located in the Netherlands, serves customers all over the world in an environment that is well recognized in both production and product with the highest design.

Moccamaster strives to produce totally reliable equipment, simple to operate, repairable, providing the maximum satisfaction with a brewing temperature between 92°C to 96°C, resulting in less sour, less bitter taste. The company contributes towards the resolution of the world's pollution and waste problems by manufacturing high quality and energy efficient "long life" products.

Moccamaster has built up a comprehensive assortment of coffee makers and coffee grinders. The family-owned company makes a distinction between appliances for private and professional use. All appliances come in a range of 18 colors and different capacity. The variants seem endless but the one commonality from these coffee masters certified by the Speciality Coffee Association (SCA), in all of their products, is a perfect cup of coffee.

MOCCAMASTER
AUTOFF

MOCCAMASTER

BRAND
German home appliances brand with over 140 years of history

ESTABLISHED
1877

PRODUCTS
Ovens, hobs, extractor hoods, refrigerators and freezers, dishwashers and automatic coffee makers

WEB
www.neff-home.com

THE STORY
The Makers at NEFF have been championing creative cooking for over 140 years. Inspiring people to explore their creativity in the kitchen has been at the heart of the NEFF brand since 1877, when Carl Neff opened his first oven factory in Bretten, Germany. Determined to focus on the person cooking rather than the technology, he developed cookers and stoves so easy to operate no recipe was too bold to attempt.

Over the decades, the range of NEFF ovens has continuously evolved to inspire new generations of cooking aficionados. They've turned their industry expertise to solving some of the basic challenges common to the main appliance found in the kitchen. Now you can cook a three-course dinner in one go without flavors intermingling. It's all thanks to their unique CircoTherm® air technology. With the unique NEFF Slide&Hide® disappearing door, nothing gets in the way of your cooking. The door fits smoothly under the oven, giving you more space and easier access.

NEFF continues to set new standards with inspired ideas, especially in its core competence areas of cooking and baking. The CircoTherm® hot air system, Slide&Hide® as the only fully retracting oven door and the effortless TwistPad® single-knob controls help get the creative juices of passionate foodies going.

Today's baking
2x vanilla 7"
4x choc 3"
2x coconut 8"

NORIAKI NARUSHIMA

BRAND
Hidden away in the Bavarian countryside, Noriaki Narushima produces his knives according to Japanese tradition

ESTABLISHED
1991

PRODUCTS
Knives

WEB
www.nori-messer.de

THE STORY
Nori, a Japanese knife maker who lives near Munich, has been making high quality knives in Germany for over 30 years. The inspiration for this Maker's calling struck him at a craftsman's exhibition in Japan where he met a famous Japanese cutlery master who shared with him the most important quality of knife making: sharpness. This simple, sublime truth brought a sharpness of vision and focus to all of Nori's subsequent creations.

He now forges his own Damascus steel using the San Mai (sandwich) technique. San Mai roughly translates to three flat layers, the core layer of steel being the hardest and the two outer layers softer. This construction allows both sharpness and flexibility.

Nori produces cooking knives, hunting knives, and switch blades. Customization in type of steel, shape, and handle material can be requested.

OTTO GOURMET

BRAND
Supplier of fine meats from butchers and farmers all around the world

ESTABLISHED
2005

PRODUCTS
Good meat, "Perfect Meat Academy", training and events

WEB
www.otto-gourmet.de

THE STORY
Otto Gourmet prides itself on making the enjoyment of quality meat at home possible. Germany's biggest online meat supplier offers products from species-appropriate and sustainable animal husbandry only. Japanese Kobe Beef, US Black Angus, Irish Hereford Prime Beef, Bellota Quality Iberico, Bison and French poultry are all sourced. This family run company offers a culinary variety which will make every gourmet's heart beat faster.

Unusual meat cuts, hearty rib eyes, delicate fillets, and mouth-watering porterhouse steaks are just a few examples that will be a delicious experience you will never forget. Customized cuts are made in house, while aging chambers refine the products to perfection – all based on customer demands.

In 2011, Otto Gourmet and its team of chefs started offering events and training sessions for professionals and private customers. General courses teach about the quality and flavor of different breeds of livestock, and for the professional, courses on how to carve and prepare whole animals.

Orders from the online store will be delivered quickly and safely to your home. Culinary delight guaranteed.

LIVAR PORK CHEEKS

Pork cheeks are part of the lower joint´s musculature and are especially well-suited for stew dishes. The Livar pig is cultured in a monastery in the Dutch province Limburg where it is free to roam and wallow in nature until it grows into maturity. Thanks to the ability to store intramuscular fat the meat is thoroughly marbled with an intensive, natural flavor.

SIMMENTALER FILET MEDAILLON

Our Simmentaler cattle, originally from Switzerland, have been bred in Traunstein, Bavaria. This beef is not only tender and succulent, but also very aromatic in its taste. The tenderloin medallion is a piece of the tenderloin center cut which makes it an absolute premium cut. For people who are not that fond of fat, the tenderloin medallion is the perfect choice.

MORGAN RANCH WAGYU 5-RIB BONE IN CHUCK STEAK

Morgan Ranch Wagyu Beef from Nebraska, USA is intensely marbled with fat, its juicy, rich flavor melts in your mouth. The Wagyu 5-Rib Bone in Chuck Steak is cut from the beef neck. It consists of the Teppanyaki Cut (or Chuck Flap) and the Chuck Roll – the most tender cuts of the neck.

BISON STRIP LOIN STEAK

From the Bison's native feeding grounds in the North American prairie, this meat has a savory, primal flavor and is leaner than poultry. In addition to a lot of vitamins, bison meat has high proportions of iron, zinc and selenium. Strip Loin Steak is a classic loin cut with its characteristic fat cap which gives the steak its juiciness.

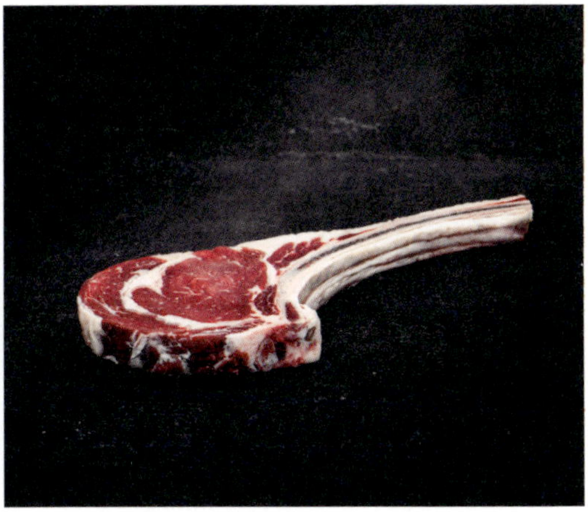

MORGAN RANCH US BEEF TOMAHAWK STEAK

The Morgan Ranch US Beef Tomahawk Steak is a ribeye on an extra-long bone. It is a majestic, succulent centrepiece for any plate. Light fat veins evenly line the reddish meat. The distinctive bone of the US Beef Tomahawk rib eye steak underlines the intense flavor of the meat. On the Morgan Ranch the cattle live in loose housing on 50 sq. km in Nebraska, USA.

KOBE WAGYU BEEF LOIN STEAK

Kobe is the beef synonymous with Japan. The term "Kobe beef" is legally protected and reserved for the thoroughbred Tajima cattle from the Japanese Hyogo prefecture. Otto Gourmet is one of the first retailers in Germany to sell Kobe beef by portions. Original Japanese Kobe Wagyu beef loin steaks are highly flavorsome and tender with the fifth taste "Umami".

GARIMORI IBÉRICO COSTILLA BELLOTA

Garimori Ibérico Costillas are marbled, succulent Ibérico spare ribs with a nutty acorn flavor. Free-range pata negra pigs with a bellota grade are fed a diet consisting of 30-40 % acorns. The pigs forage for these nuts among grass and herbs in the cork oak forests of the Spanish Dehesa. A classic for your grill and smoker.

DIETER MEIER STRIP LOIN BRANDY DRY-AGED

Dieter Meier breeds Black Angus and Hereford cattle on his Hacienda in Argentina. The Dieter Meier Strip Loin Steak is a cut from the lower back and is special because it is dry-aged 15 days in a linen sheet soaked with Spanish Carlos Primero Brandy. The result is a unique Strip Loin Steak with a light character of brandy.

UNICATE MANUFAKTUR

BRAND
Highest quality cutting boards

ESTABLISHED
1999

PRODUCTS
Cutting boards, cooking implements

WEB
www.unicate-holz.de

THE STORY
Matthias Attelmann is a passionate carpenter. He speaks to the person who wants to sign his or her living space with individuality. He believes it is not to be done loud, but with a constant, dependant persistence that is reflected in every one of his cutting boards. He appeals to the man of pleasure, who wants to feel excellent quality and attractive design under his fingers. The one who would like to be advised without having to be silent in the process itself. "Unicate" means uniqueness. Matthias takes advantage of the usability of unique natural resources and shapes them by his knowledge into the perfect cutting boards.

Unicate specializes in the production of the very highest quality products for the gastronomy and gourmet avocation. A quality that has become very popular in haute cuisine and with many an ambitious hobby cook.

BRAND
Ceramic grinders, designed and produced entirely in Japan

ESTABLISHED
1978

PRODUCTS
Coffee grinders, spice mills

WEB
www.porlexgrinders.com
www.heldbergs.com

THE STORY
Porlex is an industry leader in the design and development of ceramic products in Osaka, Japan. Known around the world for their ground breaking coffee grinders, the specialty coffee industry has set Porlex Grinders as a benchmark for quality. Their unique ceramic manufacturing technology produces the most precise ceramic burrs on the market. These ceramic conical burrs and their stainless-steel housings make for easy clean-up. They are more resilient than glass burrs, don't rust and stay sharp for much longer than steel burrs do.

Not only are these grinders designed in Japan, but they are also made to exacting standards entirely in the Porlex factory in Kagoshima, Japan. Ensuring the highest level of quality and oversight.

There have been many attempts to copy the quality of Porlex burr sets, but they are never equaled. Porlex does not provide burr sets to be used in any other grinders, so only grinders with the Porlex makers mark use authentic Porlex burrs.

ROKA / AIRSTREAM

BRAND
German-American food trailers and trucks

ESTABLISHED
1966

PRODUCTS
Food trailers and trucks

WEB
www.roka-werk.de

THE STORY

This family-owned company, which was founded in 1966, is the leading specialist in top of the line, custom vending vehicles for the food industry. Specializing in the construction and upgrading of food trucks, vending trailers and vending containers, they set new standards in the field of mobile food vending. For decades, Roka has been the embodiment of unsurpassed craftsmanship, utilization of premium materials and singular innovative prowess.

Since 2007, they have also been an authorized reseller for the Airstream brand in the European market. These celebrated US caravans are sold both as classic travel caravans and diner models tailored to the catering industry.

Roka's passion and solution-oriented work are the driving forces behind the relentless refinement of their product selection, allowing that product portfolio to grow for more than 50 years. Their success hinges on the trusting relationship they share with customers and partners. An ongoing communication with them allows Roka to offer features that are optimized for the customer's specific requirements.

The final product are trailers, food trucks and containers that give Roka customers a decisive edge over their competitors in a fiercely competitive industry.

WINGS & DRUMS

Nelles
CATERING
Est. 2016

SCANPAN

BRAND
Rooted in traditional craftmanship, Scanpan creates functional and timeless kitchenware for contemporary cooking

ESTABLISHED
1956

PRODUCTS
Cast aluminum products, kitchenware, pots, pans

WEB
www.scanpan.eu

THE STORY
With deep roots in traditional craftsmanship, SCANPAN creates functional and timeless tools for modern kitchens. This brand is constantly evolving, always at the forefront of innovation and technology, while not losing sight of their greatest achievement. Their hand cast aluminum products have formed the core of this brand since the beginning, and today they still cast all non-stick aluminum products by hand at their own factory in Denmark.

It was almost by accident that the first pressure-cast pan came into existence in 1956. Inventor G.A.L. Thorsen founded the pressure-cast steel wash, and was working on developing a pressure cooker. He eventually gave up on the project, but what was originally intended to be the lid to the pressure cooker, became the first version of a frying pan. And while one of the original aluminum presses still remains in use in SCANPAN's current production in Ryomgaard on Djursland, quality, design and functionality have been greatly developed since.

By producing all of their non-stick products in-house, they maintain complete control over all links in the value chain.

SCANPAN utilizes 100% recycled aluminum. Combined with the extended longevity of their products, the use of recycled aluminum helps reduce their carbon footprint, ensuring they make the most of the resources available. Taking traditional craftsmanship techniques as a starting point, SCANPAN goes to great lengths to ensure that every product meets strict demands, not just quality and durability but also environmental sustainability.

Liquid aluminum is poured by hand into a mold and die-cast under 250 tons of pressure. In the following processes, hot titanium is bonded to the compressed aluminum and a non-stick coating is applied and tempered. This unique casting technique ensures maximum durability and perfect heat distribution across the entire surface.

As part of their ongoing product and technology development, SCANPAN have developed a coating 100% free of PFOA and PFOS toxins, combined with a patented non-stick coating technology, making it one of the market's most resilient. The company is 100% Danish-owned, with the company's headquarters and factory located near Aarhus, Denmark. SCANPAN A/S also has offices in Finland and subsidiaries in Norway, Singapore, China and the United States and exports to more than 50 countries worldwide.

SCANPAN

YRON

BRAND
Contemporary barbecues created with traditional cast iron molding

ESTABLISHED
2017

PRODUCTS
Cast iron grills

WEB
www.yron.de

THE STORY

With a simple design and optimal performance, the Yron Grill is, in a word, beautiful. The luxurious look and feel was the brainchild of designers Frank Person and Daniel Lechowski. Applying computer graphics design to traditional methods of pouring molten iron into sand molds maximized performance and minimized waste. It's not an overstatement to say that the grill will elevate your status in the neighborhood.

Due to the production process, each grill comes out a little different, making each creation unique. For a cast iron grill, it's surprisingly light and portable, with quick assembly and disassembly. The legs are water resistant and made of oak which, over time, takes on the delicious smell of BBQ.

Since the surface of the grill is quite large, different temperate zones can be created to cook food at different speeds and heat, slow for vegetables or fast for meat. The absolute king of the German garden grills. It fits any setting, in the back yard, the balcony or garden.

With an unparalleled visual appeal and performance, this grill is in a class of its own. There's really nothing better than a sunny afternoon barbequing on the grill, and the Yron will not disappoint.

YRON

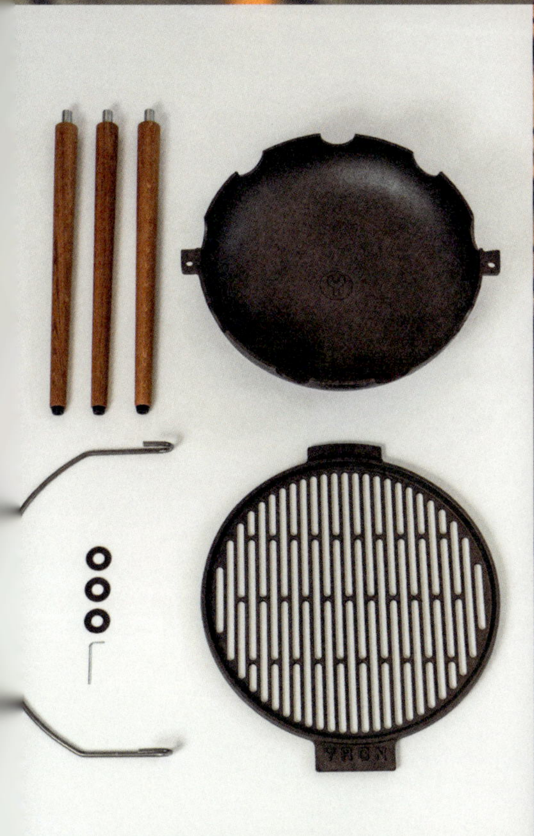
YRON

YRON

MEN OF STEEL

LARS HARMSEN

THE MAKERS BEHIND THE MAKER

Berthold Hoffmann is a Maker (see page: 170). Metal is his material, cast-iron to be precise. We wanted to know how his products are made, so he took us to the foundry Max Ginter Giesserei in Neumarkt in der Oberpfalz where they specialize in handling small pieces and individual prototypes. The principle is deceptively simple. The blast furnace heats and melts metal that is then poured into a form. Of all the manufacturing workshops, ateliers and factories we have seen, this was the most intimidating. The pungent smell, the dust in the air, and the noise were nearly overwhelming, but the heat was singular in its intensity and power!

LARS HARMSEN

LARS HARMSEN

LARS HARMSEN

LARS HARMSEN

Musterstücke u.
PROBEN
bitte hier
50x50x15
216 / 2,5

LARS HARMSEN

CHAPTER THREE

PRESERVE & SEASON

The first thing that humans kept for longer than nature intended was food. We learned how to treat our needed nutrients with certain processes and store them in ways that would ensure their longevity. These time- honored methods and implements continue to hold us in good stead.

CHILES HERMANOS

BRAND
A fraternal society for the production of hot sauces

ESTABLISHED
2015

PRODUCTS
Hot sauces

WEB
www.chileshermanos.com

THE STORY

Chiles Hermanos is truly a fraternal society for the production of craft hot sauces. Their origin story starts with founder César Gómez-Mora. After a 14 month world trip, César found himself back in his native country of Spain and immediately started missing something: hot and spicy food. His travels had taken him through many countries like Mexico, Peru and Thailand, where they traditionally serve spicy hot food, and this fire in his meals grew on his taste buds. Unfortunately, there are not a lot of hot dishes or foods in his region of Spain. They have Padrón chile peppers, piparras and cayenne peppers, but the average Spaniard of his area, is not usually thrilled to have a fiery dish set in front of him.

Beginning as a way to satisfy César's craving for burn, Chiles Hermanos quickly grew into a vehicle for delivering spicy heat to all his friends and customers. He rented a small kitchen in a restaurant and started using it during their off hours to make small batches. The head cook helped him to improve the flavors and perfect the recipes. César started selling directly and through a small shop in his local Mercado (market). His follower base grew slowly, and every month they were making bigger and bigger batches.

The Chiles Hermanos motto is "Supreme Heat". Their ultimate goal is to reach the perfect balance between heat and flavor, between pain and pleasure.

SUPREME HEAT
CHILES HERMANOS
SALSA PARA DIPPEAR PICANTE

BRAND
Legendary useful lubricant

ESTABLISHED
1904

PRODUCTS
Universal lubricant, oils, insect repellent

WEB
www.ballistol.de

THE STORY

In 1874, Friedrich Wilhelm Klever founded the 'Chemical Factory F.W. KLEVER' in Cologne. He began with the production of oils and fats based on coal and purchased a mine in order to be independent of raw material deliveries.

Around the turn of the century, the emperor sought a well-rounded oil that would maintain and preserve not just metal parts, but also wooden shanks and leather. At the same time, it was to serve as a wound oil for injuries, lacerations and bruises. In 1904, Dr Helmut Klever, the son of the company founder, who at the time was a lecturer in chemistry at the Technical University in Karlsruhe, successfully formulated this special oil. It soon became known as a "miracle oil" with surprising properties.

Produced in Germany with the original recipe for more than four generations and sold worldwide, this miracle oil is now used for everything from moving parts, gardening tools and hardware to protection of wood, metal and all kinds of vehicles.

BALLISTOL is the only universal oil to have received a "Very Good" rating when dermatologically tested.

EDELKONSERVEN

BRAND

Edelkonserven Manufaktur uses seasonal ingredients of the highest quality to create over 30 varieties of pestos

ESTABLISHED

2006

PRODUCTS

Extraordinary pesto, herb pesto, chili pesto

WEB

www.edelkonserven.at

THE STORY

In 1998, Stefan Grossauer studied agriculture and learned to relish the tastes and smells of the plants growing all around him. Six years later, he started a vegetable garden at his parent's vineyard and began selling his first preserves at farmers' markets, but after a year, he opened his own shop. After a few more years with a professional chef, Suzana Martinovic, as partner, new innovative types of preserves were developed.

The business ran so well, that more people were added to handle the demand. Cherries, herbs, nuts, and garlic were planted at the Vineyard, and the pesto preserves absolutely flew off of the shelves. Quality was and is always number one and has been maintained throughout the years of growth. Chili was introduced and many other innovative recipes were tried and customers just couldn't get enough.

In 2012, the business had to expand again and added a visitor center and tours. The taste of their pesto will have even the biggest critic singing, and the passion that goes into making it can be detected with every bite.

EXALTA

BRAND
Hand crafted chocolate made in Barcelona

ESTABLISHED
1983

PRODUCTS
Hand crafted chocolates

WEB
www.exaltachocolate.com

THE STORY

Alex and Alex are second generation owners of a little patisserie in a quaint town in the province of Barcelona. The bakery offers lots of sweets and cakes, but the Alex's wanted something to revolutionize the market.

When they met each other, they knew they had the same vision, and with the same name also, nothing could stop them!

Alex has had a long career in the patisserie world, brings sanity and balance to the task, and is very passionate about his work. Alexandra is an entrepreneur and non-conformist and drives the business with vitality and perseverance.

After getting married and giving chocolates as party favors, the positive feedback propelled them to start their revolution with chocolate. In 2017, after working for months on flavor combining and testing different products, Exalta Chocolates was born.

Exalta stands for the passion of creation, team work, social and environmental responsibility, quality ingredients, and the exclusivity of their products and services. If you can dream it, they can create it in chocolate!

"LA MEDITERRÁNEA"
CHOCOLATE BLANCO
CON ORÉGANO Y FRESA
100gr.
EXALTA HAND CRAFTED CHOCOLATE

HORBACHER MÜHLE
Spelt Semolina

Purely satisfying in its nutty taste, this is ideal in pasta or porridges and also full of important vitamins thanks to its gentle and natural production in North Rhine-Westphalia.

KELTEREI MEHLHORN
Apple Juice

This sweet and sour apple juice is traditionally made with 100% pure fruit. This morning beverage's unique aroma is comprised by all the types of apples that grow 'free range' in Saxony.

COOKAMANIA
Bacon Jam

The perfect present for every husband or father – this Berlin made jam tastes sensational. Smoked, cured bacon, honey, different spices and a little coffee form just a part of its secret recipe.

SEEDHEART
Superfood Mix Classic

This superfood, perfect for paleo diets, is a mix o linseeds, sesame and quinoa, providing you witl raw power. It is sweetened with dates and rea vanilla and manufactured in Hamburg.

KEBE LIVING
Organic Sandwich Salt

There is no better way to enjoy this salt than on freshly baked bread with butter. Naturally manufactured in Portugal and refined with selected herbs like parsley, thyme and coriander.

GARTENHAUS TESTORF
Redcurrant Jelly

This creamy fruit jelly, high in fruit content and velvety softness, is handmade from pure juice by Dorothy van Hülsen, who mainly uses fruits from her garden in Schleswig-Holstein.

NIRRA
Greek Pine Honey

Nirra makes malty and liquid honey from honeydew collected in the Greek pine woods. Its high quality is accentuated by the dark amber color, the nutritious facts and the packaging.

ZAGLER MÜSLIBÄR
Organic Honey Muesli

Nutty hemp and linseeds, combined with honey made from blossom and wholegrain oat flakes make this muesli a crunchy morning meal. All organic and created by an Austrian manufacturer.

MARTELLI
Spaghettini

Manufactured in Lari, Italy, this spaghettini's delicate yellow color is based on the traditional Martelli recipe that also involves a slightly roughened pasta surface which absorbs pasta sauce.

REISHUNGER
Basmati Rice

This traditional quality Indian Basmati grows at the bottom of the Himalayas and emits a mild earthy aroma. Reishunger cares for authentic growing conditions and a 100% sorting purity.

KEBE LIVING
Pomegranate Balsam Cream

A Berlin made vinegar containing exotic pomegranate syrup and pure red wine vinegar. Its famous for a mixture of sweet fruitiness and strong sourness and tastes fantastic in champagne.

KORONEIKI
Olive Joy Oil

Producing a high quality, extra virgin olive oil in Greece and presenting it in colorful cans, for Koroneiki, there is only one olive to press, the strong and aromatic variety of the same name.

EOMMA KIM
Korean Spicy Marinade

Made of only natural products, this authentic Korean sauce, in its elegant clay bottle, is very much the taste of high cuisine. It can be used as marinade for Bulgogi to barbecue sauce.

CHIEMGAUER GENUSSMANUFAKTUR
Organic Vegetable Stock

The Chiemgau region in Bavaria is the home of this aromatic and intense vegetable stock. It is based on carrots, parsnips and onions, contains no yeast and tastes completely natural.

KEBE LIVING
Forest Berry Mustard

The sweet taste of dark berries and the intense aroma of mustard seeds make this condiment a hot temptation. Made in Berlin, this mustard tastes creamy and goes well with cheese.

SOULSPICE
Organic Chili Peppers

These small Bird's eye chilies grow in southern India. They emit an intense smell and spicy flavor. Soulspice from Baden-Württemberg puts them in nicely designed jars for your pleasure.

HARTLS ÖL

BRAND
Fine oils with flavors ranging from poppyseed to macadamia nut

ESTABLISHED
2003

PRODUCTS
Edible oils

WEB
www.hartls-oele.at

THE STORY

In 2003, Franz Hartl founded Hartls Öl, dealing exclusively with the manufacture and distribution of high quality edible oils. In the years since, Franz has developed a few very innovative and delicate products and is a much sought-after pioneer in his chosen field. For instance, he still is the only provider of handcrafted specialities like tomato or strawberry seed oil for the international gourmet food market.

All oils are made with great care, the best raw materials and high craftsmanship. Franz appreciates delicate food and pays close attention to healthy, genuine products. The Makers at Hartls Öl advocate diversity, offering their customers the finest, small batch, unique oils while constantly working to guarantee flawless products and optimal service. Their passion for the highest quality oils is intended to provide engagement with natural oils in an autochthonous sensory experience, a real culinary adventure.

BRAND
A dark chocolatey malt-liqueur straight from Hallertau, the world-famous hops-growing region

ESTABLISHED
2014

PRODUCTS
Beer liqueur

WEB
www.heiland-liqueur.de

THE STORY
What if you could make a syrup derived from beer to use in some specialty cocktails? This is the question that started the founders of Heiland brewery on their adventure. The idea was to create a liqueur from beer that would have that powerful malty flavor. After seven months of experimenting and brewing, the final recipe was born, resulting in a complex and modern liqueur.

Heiland has a unique taste, a mixture of malt, caramel, dark fruits and a bit of chocolate. The character of the liqueur is balanced, varied, smooth and intense, and something altogether new. Building on the traditional ways of brewing from the middle ages, when monks turned brewing into an art form, Heiland has produced a novel new liqueur.

The Heiland brewery is located in the Hallertau area, one of the oldest places hops are cultivated. Heiland is completely handmade and bottled on site and could be described as “dark beer meets barrel-aged rum”. Its taste is rounded off with high quality spices like vanilla that are acquired from manufacturers known personally.

KNALLE POPKORNDITOREI

BRAND
Even popcorn can become a culinary surprise in the hands of a dedicated chef

ESTABLISHED
2016

PRODUCTS
Specialty popcorn

WEB
www.knalle.berlin

THE STORY
The Popkornditorei Knalle was founded by Lucie, Christopher and Johannes. Lucie's background is Michelin-Star gastronomy, and she's the brains behind all the brilliant Knalle recipes. Johannes is a graphic designer responsible for Knalle's look and feel online and offline. Christopher is the CEO and responsible for the company strategy and operations. The founding of Knalle happened in 2016 when Christopher and his neighbor, Johannes, approached Lucie to create their dream team, which has been producing extraordinary and completely handmade pieces of delightful popcorn in their Berlin based shop ever since.

The product range is wide, from sweet flavors like "White Chocolate Salt Pretzel" to sweet-savory like "Barbecue Smoked Almonds". In addition to regular flavors, there are seasonal flavors, for example, the summer's "Earl Grey Lime". However, nothing beats the classic, "Butterscotch Tahiti Vanilla".

Knalle is currently only sold in Germany and found exclusively in delicatessens or specialty stores.

BRAND
The German way of making chutney

ESTABLISHED
2013

PRODUCTS
Onion chutneys

WEB
www.meinmuehlenfeld.de

THE STORY
Mühlenfeld is a small shop in lower Saxony in the middle part of Germany. Mareille Willmann invented a new and tasty way to prepare onions with quality ingredients that are completely vegetarian. Raised in a family where experimenting with new food combinations was part of her daily life, Mareille started designing her own recipes at a young age.

In 2015, she decided to use local ingredients to make a mixture of oil, vinegar, beer, herbs and onions, that is not like any chutney or dip, but something new altogether called a 'Zwiebelei'. The mild cooking process keeps the onion crunch intact. The 'Zwiebelei' is sold in designer glassware with a modern look and the Mühlenfeld's logo of the stylized white onion.

With four different varieties: Red Wine Rosemary, Ginger Curry, Tomato and Allersheimer Baltic Porter Beer, every version has something special. They are perfect to flavor meat, cheese and alongside sandwiches, as well as to spice soups, use as marinades and dressings, and a whole lot more.

LES RHUMS DE CED

BRAND
Rum liqueur with fresh selected fruits and spices

ESTABLISHED
2011

PRODUCTS
Fresh fruit infused rums

WEB
www.lesrhumsdeced.fr

THE STORY
Fresh fruit infused rums are part of the historical and cultural tradition of the French West Indies. Recipes are most often handed down in the family, generation after generation. Cédric Brément grew up with the great memory of family meetings and especially the ritual of his grandfather's 'Ti Punch', a homemade recipe. Passionate about rum, he decided to improve on the way of making fruit infused rums by selecting the very best ingredients and working them in a premium craft way. After a few years of experimentation and creation, he launched Les Rhums de Ced, in 2011. This little shop of Makers is located close to French Brittany region between Nantes city and the Atlantic Ocean.

An engineer in the agrofood business, Cédric Brément pays a high-level attention to the ingredients selection and to the way his recipes are prepared. A Premium white rum from Martinique has been chosen. Part of the rum is matured in oak barrels that have been used in different famous French wine regions to experiment in a new way of rum tasting. Fresh perfectly ripened fruits arrive every morning to the workshop and they are immediately cut and bottled to preserve their natural flavors. All fruits are cut with a knife and bottled by hand with a Bourbon vanilla pod from Madagascar.

BRAND
Honey from next door

ESTABLISHED
2014

PRODUCTS
Honey

WEB
www.nearbees.de

THE STORY

Environmental protection, sustainability, local and conscious consumption, this is the vision of nearBees. With more beekeepers, we have more bees, and with more bees we have more diversity in nature! Based on this philosophy, nearBees connects consumers and beekeepers via an online marketplace. Consumers can easily find their local beekeepers and purchase honey directly from them, supporting local businesses instead of reaching for cheap, imported options in the supermarket.

nearBees provides a win-win situation for all, by offering an easy way of buying and selling local honey: transparent, regionally sourced, and an avenue to sell at fair prices for beekeepers and to buy natural raw honey for honey lovers. In this way, everyone can do their share in stopping the decline of bees – for a blooming and livable neighborhood.

NEARBEES

nearBees
HONIG VON NEBENAN

BRAND

Nuts and seeds are pressed into fine oils in this Berlin-based oil mill

ESTABLISHED

2012

PRODUCTS

Cold pressed oils

WEB

www.oelwerk.de

THE STORY

At Ölwerk, indulgence is their passion. Once you've tasted these freshly pressed, organic cooking oils, you will not be able to do without them.

After pressing, the oil is not treated any further. It consists only of the nuts or seeds; therefore it is important that these raw materials are of the best quality. To ensure this, Ölwerk process only organically grown, nuts and seeds. Once pressed, there is a resting phase in which suspended particles settle. After this decanting, the oil is bottled and sent right out to the customer. They only press as much oil as they have orders to sell directly.

Seasoning and cooking oil is traditionally a regional product. In the 19th century there were 4,000 oil mills in Germany, which marketed their oils regionally. Due to industrialization and the associated centralization there is little left of that tradition and business.

Although Berlin has a distinct tradition in the production of linseed oil, there was no oil mill for seeds and kernels. Ölwerk changed that, opening its doors in 2012.

125 mL 1/4 TEASPOON

BRAND
Spice Mountain offers a range of spices from around the world, specializing in handmade spice blends

ESTABLISHED
2008

PRODUCTS
Herbs and spices

WEB
www.spicemountain.co.uk

THE STORY
Spice Mountain was born when Magali Russie and Matt Norris visited a spice plantation in Goa, India almost 10 years ago. They returned to London inspired by what they had seen, and began selling spices from a small market stall. Since then, Spice Mountain has grown into a company with retail outlets in two of London's best-known markets, and also a thriving online shop – the go-to place when you are looking for the very best in spices, herbs and associated products.

Spice Mountain's success has been built on a commitment to sourcing and working with only the best in spices, and to this end suppliers are chosen with great care to ensure that everything used is produced ethically and sustainably. The company's flagship retail outlet is in London's iconic Borough Market.

The range of products supplied is incredibly wide and genuinely global, ranging from basic herbs and spices to exotic rarities. Spice Mountain specializes in handmade spice blends using authentic recipes from around the world, and prides itself on blending to the peak of perfection every time, a process ensured by the fact that founder Magali still makes the blends herself.

Spice Mountain's philosophy is simple – to celebrate spice, and to share the dazzling array of colors, scents and flavors involved by offering hundreds of ways to sprinkle a bit of magic into the cooking experience.

SPIRIT OF OAXACA

FER ARCE

TOUGH CRAFTED MOUNTAIN MEZCAL

The plateau of Oaxaca hosts the village of San Juan del Río. This village has a century old tradition in growing and harvesting agave. At 1,500 m above sea level and influenced by the Pacific, the location provides perfect conditions for the plants. It is a tough job harvesting, cooking and distilling the juices into Mezcal. Photographer Fer Arce's pictures bear witness to the sweaty task.

FER ARCE

FER ARCE

FER ARCE

FER ARCE

FER ARCE

FER ARCE

FER ARCE

FER ARCE

FER ARCE

CHAPTER FOUR

HYDRATE & DISTILL

Water is life, but more than that, it is an ambassador for whatever other element we want to take in. From coffee to whisky, tea to schnapps, we have been creating and loving libations of varying kinds from the moment we could appreciate imbibing something to not just quench our thirst.

AQUA MONACO

BRAND
Purest natural mineral water and contemporary mixers from ancient glacial plains around Munich Germany

ESTABLISHED
2011

PRODUCTS
Mineral water, artisan beverages, botanical drinks

WEB
www.aquamonaco.com

THE STORY
From Aqua Monaco come beverages for special venues and sophisticated gastronomy. A group of young makers, driven by an unmistakable desire for outstanding quality and contemporary design became fuelled by their curiosity and passion. This forward-thinking shop started, and rightly so, with the most important raw material for drinks: the water.

Choosing to use a natural mineral water from deep beneath the ancient glacial plains around Munich as the foundation for all of their sodas, tonics and soft drinks, leaves each of Aqua Monaco's products with an unmistakable geo-fingerprint on the taste buds. This water is particularly pure, gently mineralized and incredibly smooth. Still or sparkling, the taste leaves its unique, clean signature on all of their liquid libations.

The Aqua Monaco mixers combine natural ingredients of exceptional origin into innovative recipes, distilling one of a kind creations, making any drink using them, stand out among the crowd. These, as well as all of their fluid fancies are filled exclusively in glass bottles with the unmistakable Aqua Monaco design.

MONACO
Mineralwasser
still
AQUA

BELSAZAR

BRAND
Where Berlin meets the Black Forest

ESTABLISHED
2014

PRODUCTS
Vermouth

WEB
www.belsazar.com

THE STORY

What happens when you combine burning passion for regional wines from Baden, a rich distillery culture and a sense of creativity nurtured on the streets of Berlin? In this case: A wine aperitif in a class of its own.

Based on hand-selected, regionally-produced wines from the regions of Markgräflerland and Kaiserstuhl in South Baden, the founders of Belsazar translated their love for high quality regional ingredients and wines into a unique German vermouth. With distinctive flavor profiles that stem from an array of herbs, spices, blossoms and peels – a different composition for each variety. Then a hint of fruit brandy from the infamous Schladerer distillery, family-operated since 1844, gives Belsazar its finishing touch.

With a fresh take on a vintage bar staple, Belsazar offers four different varieties of vermouth. Within those you can find classic vermouth types such as Sweet, Red and Dry but also modern varieties such as Rosé.
This makes Belsazar an ideal partner in classic cocktails such as the Martini, the Manhattan or the Negroni but also offers new takes on low alcohol by volume aperitif drinks. Due to their unique full-bodied flavor profiles, the Belsazar varieties will work great as well-balanced long drinks with tonic water and ginger ale.

BELSAZAR
VERMOUTH
ROSÉ

BRAND
The world's first mountain pine soda

ESTABLISHED
2017

PRODUCTS
Custom beverage

WEB
www.bergbrause.com

THE STORY

Bergbrause is the world's first mountain pine soda. Traditionally produced from natural ingredients, this liquid slice of crisp mountain air is the creation of Hannes Drensler. The man behind Bergbrause came up with the idea while hiking in the Alps. His dream was to make a soda that transported you instantly to that unique mountain landscape – so fresh that as you drink it you can almost smell the Alpine air. The scent of the mountain pine is typical of the entire region. In fact, mountain pine oil has been used for generations as a natural home remedy. Hannes wanted to find a way to bring this characteristic olfactory signature to a beverage. The challenge was to capture the unique flavors contained in the twigs and needles of the tree, in just the right proportions. He found the perfect partner to help him master this in a fruit juice production facility in Berchtesgadener Land, Bavaria, Germany. The contrast between the bitter pine and the fruity organic apple juice is what makes the taste of Bergbrause so special.

BRAND
Premium herbal tea made from a variety of Greek blends

ESTABLISHED
2017

PRODUCTS
Premium teas

WEB
www.grektea.com

THE STORY

The makers at GREK wanted to recreate for others that magical moment they first experienced drinking a traditional Greek herbal infusion. They set out to develop a range of premium herbal teas consisting of Greek blends, based on traditional recipes, along with the more familiar flavors such as peppermint and chamomile.

They provide the best herbal teas Greece has to offer by selecting only the highest quality organic herbs from all over Greece and working side-by-side with small-scale farmers. The ingredients are carefully selected, not only for their individual quality but also for their complementary properties when blended. By carefully cultivating, harvesting and preserving the purest plants, they ethically source products that are more than tasty; they're good for you.

GREK launched in April 2017, bringing their new range of premium herbal teas to the UK market. This brand steers clear of Greek clichés and has instead taken inspiration from the Xysta tiles of Pyrgi, Chios. The result is a powerful, unique and memorable range of products with a contemporary feel. The packaging is of high quality and each tea flavor comes with a little booklet featuring a unique short story and charming illustrations.

BIRRA KARMA LEMON ALE

With the addition of coriander, the golden Lemon Ale resembles more of a Belgian Witbeer. But the differences can be savored on the tongue. The fruity taste is noticeable, while the coriander produces a slightly harsh note at the beginning, finally finishing with a pleasant sweetness.

ESPIGA BLONDE ALE

The Catalan brewery Espiga gave us a true fruit bomb in their blonde ale. The cloudy-golden ale smells beautifully of tropical fruits, white grapes and watermelons. It does not disappoint in the taste either: the tropical fruits persist while the finish is long-lasting and bitter.

ST. ERHARD ORIGINAL

What distinguishes an original is its distinctive character. These noble roast malt aromas transport restrained sweetness, while the hops suggest a balanced bitter note. Overall the St. Erhard original is a restrained contemporary - not too strong, not too tart.

LEHE AND ST. ERHARD ENDSPIEL

From breweries Lehe and St. Erhard, comes a barleywine like no other. Flavors of dried fruits and caramel rise in the nose to be followed by a pleasantly sweet and powerful finish. The Endspiel of Lehe and St. Erhard definitely leaves a lasting impression that makes you want more.

LEHE HECKLE

"Heckle" in Estonian loosely translates to "interjection". With a strong fruit aroma and balanced notes of pineapple and mango that do not escape any nose, the taste of the Heckle does not fight to compare. This beer is a balanced and voluminous body, fully engaging the senses of the drinker.

HOPFEREI HERTRICH VETO SCHOKOBÄR

Oat flakes are used to add more structure to this beer, rather than conventional stouts. The nose is assailed with roasted malt, chocolate, delicate licorice and fine coffee aromas that follow closely to the tongue. The delicate malt sweetness completes the chocolate-coffee-like taste.

SAKIŠKIU ALUS CHIPOTLE PORTER

he fire of this black porter jumps at you when ouring: roasted aromas of licorice and chocolate lay with fresh pepper and a touch of smoke. This ittersweet and a spicy libation is the collaborative rink from the brewery Sakiskiu Alus and the Mexican restaurant Soda de Pancho.

SEPTEM 8TH DAY IPA

Characterized by the aromas of tangerine, citrus, lychee, mango and peach, Septem's 8th day IPA impresses with a powerful 7.0% alcohol, harmonizing well with its rich aromatic flavor. The dark orange IPA with notes of pine, mango and piquant aromas, is an explosive combination.

BRAND
Bulleit Frontier Whiskey is a family of high-rye, award winning whiskeys with spice and complexity

ESTABLISHED
1987

PRODUCTS
Whiskey

WEB
www.bulleit.com

THE STORY
Today Bulleit Bourbon is still produced according to a traditional recipe, which is particularly aromatic, dry and clear thanks to a high rye content as well as the pure Kentucky limestone filtered water that are used for this distinctive whiskey. Apart from Bulleit Bourbon and Bulleit Rye, their new premium whiskey is exclusively bottled from a selected reserve. These stocks of extra strong burnt whiskey barrels from white oaks were stored with a minimum maturation period of 10 years. The result of the extra-long maturity reveals itself in the depth, complexity and incredible gentleness of the Bulleit Bourbon 10-Year-Old. These attributes, in addition to the high rye content, represent the character for which Bulleit is known worldwide.

Between 1830 and 1860, the restaurant owner Augustus Bulleit developed this amazing bourbon recipe. One fateful day, while transporting his barrels of bourbon from Kentucky to New Orleans, Augustus Bulleit vanished. What happened is still unknown, and his creation nearly disappeared into history along with him. The whiskey fell into oblivion until Tom Bulleit founded the Bulleit Distilling Company in 1987 and revived the 150-year-old heritage of his great-great-grandfather. Today, we're not the only ones who are glad he did.

BULLEIT
FRONTIER
WHISKEY
PLEASE DRINK RESPONSIBLY

BULLEIT
BOURBON
FRONTIER
WHISKEY
AGED 10 YEARS
BULLEIT BOURBON
FRONTIER WHISKEY
KENTUCKY STRAIGHT BOURBON WHISKEY

CAPOVILLA

BRAND
A small distillery from Italy's Veneto region

ESTABLISHED
1986

PRODUCTS
Fruit liquor

WEB
www.capovilladistillati.it

THE STORY

In 1986, after years of experimenting, Vittorio 'Gianni' Capovilla developed his own method of distillation which combined a double boiler with steam, a slower processing time, and used lower temperatures and less pressure. Capo started distilling any fruit he could find, making quality liquor that was never seen before. Only the freshest cherries, plums, apples, elderberry, rowanberry, cornelian, and others are used in his process. He's been producing "bottled excellence" since then, with passion and extraordinary ingredients. Most are distilled to 40 % alcohol. The Capovilla liquors are bottled in-house with nondescript bottles using handmade labels, showing a logo of a tree with many kinds of fruit.

In 1995, Capo entered the Austrian "Destillata", a contest normally dominated by Austrians. He won second place overall with five golds, two silvers, and four bronze. An Italian has never won with such high ranking, and this really shook up the world of distillery.

Capo had other radical ideas. One was to distil beer, and after a long battle with bureaucracy and experimenting with different beers, he finally distilled and aged a blond, double malt into a complex liquor that pairs well with chocolate or cigars.

Capo also founded Rhum Rhum PMG with his friend Luca Gargano, which distils sugar cane on the island of Marie-Galante in the Caribbean. White Rhum Rhum Agricole is distilled using fresh sugar cane juice, pressed once, like fine virgin olive oil. Capo plans to expand in this area and who knows what he will produce in the future.

CAPOVILLA
DISTILLATI

DAHLMANN CATERING

BRAND
Taylor-made catering for exclusive events all across Europe

ESTABLISHED
2007

PRODUCTS
Individualized catered events, juice, syrup, spirits

WEB
www.dahlmann-catering.de

THE STORY
With his ambitious drive for creativity and individuality, Uli Dahlmann has grown Dahlmann Catering into one of the top catering businesses in Germany. He doesn't accept any compromises when it comes to quality and offers only the finest products, many of them homegrown on the family's own orchard in Franconia or supplied by organic farmers from around rural Bavaria. The orchard has about 1,000 fruit trees that produce apples, pears, cherries, Mirabelle plums and other traditional fruits which are used for making juice, syrup and schnapps.

The company's philosophy goes far beyond that of traditional event catering. CEO and owner Uli Dahlmann and his skilled team of experienced professionals aim to create lasting memories for the people who attend his events. Clients will not only experience the finest dining, but a level of service that surpasses most Michelin starred restaurants. Moreover, Dahlmann Catering tailors each event to the needs of their individual clients – be it a classy private dinner in a sophisticated home or a glamorous fashion show.

Once you experience a Dahlmann Catering event, no other will compare.

APFELBRAND

BRAND
Made in the UK using flavor pairing techniques to create drinks with no artificial flavoring, coloring or preservatives

ESTABLISHED
2015

PRODUCTS
Mixers

WEB
www.doubledutchdrinks.com

THE STORY
Every bottle of Double Dutch is made entirely in the UK using flavor pairing techniques in order to create amazing drinks with no artificial flavoring, coloring or preservatives. Only natural ingredients are blended with the highest quality spring water from the North of England to create crisp, fresh and innovative mixers for refined palates.

From the world's best mixologists to those who prefer their G&T on the sofa at home, Double Dutch mixers have developed complex flavors that augment spirits, satisfying aromas that enhance each drink and superlative quality and smoothness to delight the palate.

A mixer should elevate the spirit, spurring one to experience richer, deeper and more exciting flavors. It should reward the drinker with sensations that improve. Rather than diluting a chosen drink – it should bring a spirit to life.

Great cooks and food creators have always had an intuitive sense of which flavors work together. Molecular gastronomy is an increasingly utilized science that explores which foods have similar flavor balances to reward the palate. This is the philosophy behind Double Dutch – drawing on scientific research, sourcing the finest ingredients and combining art with craft to create soft drinks that intensify the flavor of spirits.

DOUBLE DUTCH
DUTCH

Zacapa
DOUBLE DUTCH

DRAFT BROTHERS

BRAND
Ten spices and herbs give this gin a distinct flavor of juniper complemented by classic herbs

ESTABLISHED
2015

PRODUCTS
Gin

WEB
www.draft-brothers.ch

THE STORY
In 2015, the Draft Brothers finally decided to professionalize their passion for Spirits. Beni and Sam met in oenology school, where they not only trained their palate, but also laid a solid foundation for things to come. After spending a couple of years in the wine industry, the friends found a lot of traditional spirits had become industrialized and figured it was time for a change.

Using their trained palates as a guide, they started producing handmade, finely crafted spirits of the highest quality possible.

Every ingredient they use is painstakingly selected, distilled separately and blended to unwavering standards. In the early days at Draft Brothers, gin production wasn't a priority. But late one fall with a long winter approaching and an equally long wait before their next brew would be ready, they started experimenting.

After a year of trial and error the Draft Brothers emerged with a recipe for an old school gin. A gin where juniper flavors take center stage, backed by an orchestra of cardamom, pepper and floral notes.

BRAND
Exquisitely crafted, botanically brewed beverages

ESTABLISHED
1905

PRODUCTS
Artisanal beverages, botanical drinks

WEB
www.fentimans.de

THE STORY

In 1905 in the small Yorkshire town of Cleckheaton, Thomas Fentiman, an iron puddler by trade, provided a cash loan to a fellow tradesman. As security, Mr Fentiman received what would turn out to be an extremely valuable deposit – a recipe for Botanically Brewed Ginger Beer. The loan was never re-paid and Thomas began using the recipe to produce his ginger beer delivering it door-to-door in a horse and cart. His beer was stored in handmade stone jars affectionately known as 'grey hens' which were stamped with an image, remaining on the bottles to this day, of his pet dog Fearless.

In his original recipe, the finest ginger root was bruised, crushed, milled, then placed into copper steam jacketed pans with spring water and gradually heated, releasing a sediment and botanical extract. This liquid was then filtered into a wooden vat with the finest herbs, natural flavorings, sugar and brewer's yeast, thoroughly stirred in. After fermentation, the live ginger beer was then decanted from the vat into the iconic stone jars.

Fentimans ginger beer became very popular and was soon followed by other beverages. As the family business grew, to keep the workforce local, several brewing and production facilities were used in the North of England.

The world has changed a lot over the 100 years since Thomas Fentiman founded the business but their commitment to quality, natural ingredients and innovation remains as strong as ever. Today, Fentimans Botanically Brewed Beverages is the only botanical brewer in the UK.

FENTIMANS
BOTANICALLY
SERVE CHILLED
FENTIMANS
EST. 1905
BOTANICALLY
BREWED
TRADITIONAL
GINGER
BEER
UPEND BEFORE POURING
Fermented Botanical Ginger
Drink with Herbal Extracts
275ml e
Best before: see neck

BOTANICALLY
SERVE CHILLED
FENTIMANS
EST° 1905
BOTANICALLY
BREWED
NATURAL
PINK GRAPEFRUIT
TONIC WATER
Ruby Red Grapefruit and Quinine
Mix with Gin or Vodka for a
refreshing and zesty long drink
500ml e
BEST BEFORE: SEE NECK

FREIMEISTER

BRAND
This collective of small independent distilleries offers a wide range of fine spirits distributed under a common label

ESTABLISHED
2016

PRODUCTS
Craft spirits, whiskey, gin, herbal bitters, vodka, liqueur, brandy

WEB
www.freimeisterkollektiv.de

THE STORY
By adhering to a high standard of ethics called the Freimeister Code, the FMK offers the highest quality hand crafted spirits. They work with a collection of small, artisanal distilleries, bartenders, and experts who live by that same code. The goal is to take back the production of spirits from big business and make it fair, fresh and contemporary. The FMK products are available at fair wholesale prices for both consumer and industry alike without any common industry markups.

The FMK offers a curated selection of the highest quality spirits like amaranth vodka, white dog, gin and herbal bitters, sourced straight from the producer. Each can be sampled straight up or enjoyed mixed in a cocktail.

The Freimeisterkollektiv prides itself in being transparent with the motto that everyone should know what they're drinking and from where the ingredients came. All FMK products are free from artificial flavors and additives, yeast and GMOs, and the manufacturing processes use sustainable raw materials.

FREIMEISTER KOLLEKTIV
AMARO 212
KRÄUTERLIKÖR
THOMAS NEUBERT
Gutsbrennerei Zinzow
Boldekow / Vorpommern
BITTER
LINGER
BODY
FLORAL
CITRUS
HERBAL
FRUITY
Freimeisterkollektiv GmbH
10719 Berlin / Germany
freimeisterkollektiv.de
ROSMARIN
GEIST
MAHEMBE 336
KAFFEELIKÖR
JOSEF FARTHOFER
Destillerie Farthofer
RALF RÜLLER
The Barn Coffee
20%VOL
50 ML
KRAUSE-
MINZE
GEIST
GEORG HIEBL
Destillerie
Georg Hiebl
Reichhub
Mostviertel
40%VOL
50 ML
SALBEI
GEIST
GEORG HIEBL
Destillerie
Georg Hiebl
Reichhub
Mostviertel
40%VOL
50 ML

BRAND
Fine beer specialities brewed with hops from their own farm

ESTABLISHED
1995

PRODUCTS
Beer, beer jelly, beer mustard, black ale, pale ale, IPA, hops bitters and fruity apple liqueur

WEB
www.hopfengut.de

THE STORY
The Hopfengut No20 combines a hops farm, brewery, museum, shop and restaurant all under one roof. Here, the team supplies the finest quality hops with the best aroma for national and international breweries.

The Makers from Hopfengut No20 are passionate about brewing beer and producing exceptional specialties for connoisseurs.Beer produced here strictly follows the German purity law of 1516.

A visit to the estate in September is a must do. The harvest time brings about the rich smells of ripe hops at a beautifully mild time of year. You can experience how beer is made by unique state-of-the-art agricultural technology, as well as old style manual labor. Make it a day, have lunch or dinner in the restaurant and sample the products of the brewery.

In the Hopfengut No20 shop, selected hops specialties can be purchased, such as hops distillates, hops tea or simply good beer. In the spirit of Hopfengut, all good.

HOPFENGUT
№20
DIE BRAUEREI

RIEGELE

BRAND
This family-owned Bavarian brewery does not rest on its centuries-old tradition

ESTABLISHED
1386

PRODUCTS
Craft beer

WEB
www.riegele.de

THE STORY
Riegele brewers are passionate about their art of craft beer brewing. World Champion Beer Sommelier Sebastian Priller-Riegele's journeys through the world's top-brewing nations are the inspiration for Riegele's hand-picked selection of unique beers. The range of colors, from amber to chestnut red to deep black, and tastes, from citrus to rich chocolate, exhibit unmatched quality.

With five malts, the Noctus 100 Imperial Stout offers the black, English antithesis. Auris 19 and 20 Ator use brewing styles that originated in Bavaria – one golden crisp and one dark, but always strong in malt. The Anglo-Saxon colonial period with its great tradition of Indian pale ales is resurrected in the Simcoe 3. Augustus 8 is a fruity tribute to Emperor Augustus, the founder of the Roman city of Augsburg, the home of the brewery. Each beer is unique in its own way and represents the unparalleled quality that has become the hallmark of Riegele.

Dating back to 1386, Riegele is recognized and admired the world over as one of the oldest and finest family owned Bavarian breweries. Passion, dedication, and knowledge are the three ingredients required to produce beers of exceptional craftsmanship and artistry. These traits have been passed down from generation to generation in the Riegele family.

BRAND

A young beer brand from Munich using only organic ingredients to brew beers inspired by the city's spirit

ESTABLISHED

2015

PRODUCTS

Beer

WEB

www.isar-bier.de

THE STORY

Starting a new beer brand in Munich, the very capital of beers, might not seem like the best idea, but Maximilian Valentin had a vision and, with Riedenburg Brauhaus, he found a perfect collaboration to bring his brew to reality. The storied brew-house he'd partnered with has the same comprehensive ecological conviction, characterized by respect for man and nature, that the ISAR Beer company has. For over 20 years, they have brewed only organic, making them the perfect partner to bring a great new Bavarian beer to life!

Different to the established Munich beers, ISAR is brewed with the finest organic aroma hops from the Hallertau. Easy to drink, not too heavy but still with a personality. The best beer for a summer evening at that famous Munich river. That is the essence of ISAR Bier.

Since 2015 you can find the "ISAR Hopfengestopftes Helles" – an unfiltered and dry hopped version of the traditional Helles – in organic markets, good liquor stores and fine bars all around Bavaria. It is crafted purely out of organic ingredients. Great raw materials, great craftsmanship, great beer.

ISAR BIER

MEZCAL ATENCO

BRAND
Traditional mezcal experiences a renaissance, and Atenco is one of the distilleries responsible for it

ESTABLISHED
2015

PRODUCTS
Handcrafted mezcal

WEB
www.mezcal-atenco.com

THE STORY
Atenco loosely translates to 'the place on the edge of the water' and refers to San Juan del Río, a small town in the mountains of Oaxaca and the birthplace of traditional Mezcal. Produced with knowledge and experience for over three generations, it is handcrafted by Don Rodolfo's family and their faithful horse, Colorado.

Mezcal Atenco is made of Agave, a plant that is allowed to grow and mature at least eight years before being harvested. It is then crushed with a stone mill, cooked in an earth oven covered with banana leaves and fermented in pine tubs with fresh spring water from the mountains. This process has not changed for centuries and the results are copper pot double distilled 'Atenco Espadín' and the triple distilled 'Atenco Pechuga'. Packaged in 100% recycled, hand-blown glass bottles, made specifically for these unique libations, with a smooth, smoky and complex flavor, Mezcal Atenco embodies the tradition and spirit of the Agave plant and those who harvest it.

NARDINI

BRAND
Nardini represents tradition and constant evolution, producing extraordinary spirits

ESTABLISHED
1779

PRODUCTS
Grappa, liqueurs

WEB
www.nardini.it

THE STORY
Ditta Bortolo Nardini is an iconic brand based near Venice in Bassano del Grappa, a name synonymous with grappa. Founded in 1779, Nardini is the oldest grappa distillery in Italy and their legacy continues to live on, now in its seventh generation.

On the long road that brought Nardini to the forefront, the family has always known the importance of staying true to their roots, traditions and original recipes while keeping an eye on the future.

The master distiller practices the ancient art of blending and tasting, combined with the most advanced production systems, which allows Nardini to maintain their unique characteristics while retaining a position of leadership in the world of grappa. Nardini also produces exclusive aperitifs, Amari, and liqueurs, sold in Italy and around the world.

Each Nardini site in Bassano maintains the same philosophy and has its own intrinsic value. There is a historically-listed grapperia, an ultra-modern lab, a visitor's center designed by world class architect Massimiliano Fuksas, and the new Garage Nardini which is a modern take on a typical osteria serving products from the Veneto region.

Nardini represents tradition and constant evolution, producing extraordinary spirits.

PREMIATA DISTILLERIA
DITTA BORTOLO NARDINI
BASSANO VENETO

BASSANO
AL PONTE

NORD-T

BRAND
Tea from Finland that captures the flavors of Scandinavia's beautiful nature

ESTABLISHED
2014

PRODUCTS
Premium tea

WEB
www.nord-t.com

THE STORY
Childhood friends Mirka and Nina have known each other since Sunday school. Over breakfast one morning with a mutual friend, Mirka launched into a long monologue about how bad her tea was and Nina was struck by an idea. Why not make tea the way it was made originally, from natural ingredients? After that initial inspiration, Nord-T was founded and since then, the two life-long friends have spent more time together than with their respective boyfriends

They started their adventure hand assembling their mixes and shipments in Helsinki, but soon moved to their "little tea factory" as they like to call it, to the middle of Finnish countryside, in Kihniö. Since most of their Finnish ingredients are from Lapland or central Finland, Kihniö is the perfect place to be. Enabling them to get ingredients both straight from forests, like lingonberries and birch leaves, as well as from farms, like rhubarb and strawberries.

Their secret recipe is that they actually have no hidden ingredients. Nord-T is quite proud to say that all of their raw materials are grown by the most skillful farmer of them all - Mother Nature. They use no additives, absolutely no natural or artificial aromas, and rely just on nature's own tea leaves, berries and herbs. A more natural option for tea drinkers.

Moose on the Loose
Loose leaf black tea with blackcurrant, blackcurrant leaf and red clover
No additives, no oils. It's tea with arctic superpowers!
NORD-T
Helsinki
Loose leaf green tea Organic En Shi Yu Lu
No additives, no oils. It's tea with arctic superpowers!
NORD-T

BRAND
Fresh spring water from the mountains of Alto Adige

ESTABLISHED
1957

PRODUCTS
Mineral water

WEB
www.acquaplose.com

THE STORY
Our story starts 60 years ago on Mount Plose, from a spring 1,870 meters high in the uncontaminated mountains of South Tyrol, with the discovery of a water now recognized as one of the best mineral waters in the world. Acqua Plose has inherited a lot from those untainted peaks, which were declared a UNESCO World Heritage site. They have inherited a great love of the land and its fruits, the genuineness, the pureness, and the respect. You will find that inheritance in every single drop of Plose products.

That water, rich in its uncommon mineral properties, has inspired more equally unique products. The BioPlose 100% organic fruit juices, Tea Collection of iced teas and Plose Vintage fizzy drinks are all made according to ancient alpine recipes. All bottled in glass, all genuine and all born of the same mountains that remind us every day of the importance of safeguarding the most precious thing we have: the goodness of nature.

All bottled by Fonte Plose Spa, this company is involved in every aspect of bottling and does so according to the highest standards of environmental sustainability, and respect for the land on which it operates, including local values and local traditions. This philosophy has led to the very strictest observation of rules of hygiene dedicated to preserving the exceptional natural qualities of their products.

PLOSE
FRIZZANTE
ACQUA MINERALE NATURALE
NATÜRLICHES MINERALWASSER
PLOSE
FRIZZANTE

BRAND
Italy's first single malt, aged in vintage grappa barrels

ESTABLISHED
1947

PRODUCTS
Whisky, spirits

WEB
www.psenner.com

THE STORY
Ludwig Psenner was born in 1920 in the South Tyrolean wine village of Tramin. At an early age, he threw himself into the task of producing quality brandy - known in Italy as grappa - and in 1947, founded his Psenner distillery. Eventually deciding to apply his skills to the distillation of fruit brandies, Psenner supplemented their range of grappa with Williams pear brandy. He was the first to do so in Italy, and the product became an immediate success. Having conquered the local market, Psenner became one of Europe's leading producers of Williams pear brandy.

Always innovating, they next created the new concept of positioning bottles on pear trees, making the fruit grow directly into each and then filling it with Williams pear brandy. The "pear in the bottle" has now become ubiquitous worldwide, although others generally insert the pear from the bottom of the bottle then close it from there. Psenner has never been attracted by such shortcuts and continues to position bottles over embryonic pears in the company's own orchards.

That same burning thirst for experimentation has driven Psenner to begin maturing their first whisky in their own ex-grappa casks. Producing a uniquely rich and intensely flavored libation. Processing their fruit, using cutting-edge distilling methods, all honed by decades of experience, the Psenner distillery naturally established itself among the elite of the world's producers, confirmed by awards and the praise heaped upon its products by international experts.

GÖPPINGEN

PSENNER
ESTABLISHED IN 1947
RÈTICO
FINEST ITALIAN SINGLE MALT
WHISKY
AGED IN GRAPPA & SHERRY CASKS
Limited Edition
DISTILLED AND BOTTLED
IN SOUTH TYROL
BY THE PSENNER DISTILLERY

REDBREAST

BRAND
One of the few Irish Single Pot Still Whiskeys still available today

ESTABLISHED
1857

PRODUCTS
Single Pot Still Irish Whiskey

WEB
www.redbreastwhiskey.com

THE STORY
Redbreast whiskey takes its name from the small 'Robin Redbreast' bird, originating in 1903 with the then chairman of Gilbeys Wines & Spirits Import Company, who was an avid bird fancier.

The custom of that era was for distilleries to sell their whiskey to 'bonders,' like Gilbeys, who were in the business of importing fortified wines such as sherry and port. With ample supplies of empty casks in which to mature new whiskeys, the wine merchants would produce these whiskeys under bond, hence the term 'whiskey bonder'. These were then marketed under the merchant's own name, for example, Redbreast.

Of the many brands produced and marketed in this fashion, Redbreast was one of the very few bonder brands that endured, owing to reputation and a loyal following. In 1991, Redbreast was re-introduced by Irish Distillers Limited, after an absence of almost 10 years, and was given a thorough makeover. Different whiskeys were produced after that including: Redbreast 12, Redbreast 15, Redbreast 21, Redbreast Mano a Lamh (a limited exclusive edition) and the latest Redbreast Lustau Sherry Finish Edition.

Today Redbreast is the largest selling Single Pot Still Irish Whiskey in the world and is considered the definitive expression of the traditional Irish whiskey.

BRAND

Germany's first rye whisky distillery produced right in Brandenburg's grain growing region

ESTABLISHED

2016

PRODUCTS

Whisky

WEB

www.stork-club-whisky.com

THE STORY

Sixty kilometers south of Berlin, the idyllic Spreewald river delta and wetland extend into Brandenburg. Here, amidst a protected landscape surrounded by rye fields, lies the Spreewald distillery. It was founded in 2003 and has been run by Steffen Lohr, Sebastian Brack and Bastian Heuser since 2016.

Talking about whisky, people usually think Scotland or America, nowadays even Japan. But Whisky from Berlin-Brandenburg or the Spreewald? The area offers a genuine microclimate for aging whisky combining humidity with hot almost tropic summers and harsh cold winters. In addition, Brandenburg is Germany's bread basket with plenty of fine water around and a rich history in distilling grain. In the middle of the so-called "Rye Belt", which stretches from western France to the Ukraine and all the way through the northern part of central Europe, the rye grain grows exceptionally well and Brandenburg is the biggest cultivation region of it in Europe. Even popular US distilleries such as Wild Turkey buy their rye in Brandenburg because of its exceptional quality. This history and the rich abundance of rye around them, inspired the three friends to produce Germany's first Rye Whiskey brand, called Stork Club. In 2017, Spreewood distillers installed a new Kothe still, new mash tuns, fermentation tanks and a grain mill to produce their whisky strictly from grain to glass on premises.

GUGGENHEIMER

BRAND
Slow roasted coffee made of selected coffee beans

ESTABLISHED
1932

PRODUCTS
Coffee

WEB
www.guggenheimer.coffee

THE STORY

Good things take time. The founders of Guggenheimer Coffee know the slower, the better. Keeping to this ethic, they produce a softer, richer flavored bean.

Industrially roasted beans are often roasted at very high temperatures for only 2 to 5 minutes. The resulting coffee is usually too bitter and too sour. Guggenheimer Coffee is roasted at low temperatures for up to 30 minutes per batch. Undesirable acids and bitter tastes are degraded with this slow roasting method. To add to the great taste experience, the Guggenheimer family carefully selects raw beans from the best coffee regions of the world.

With every sip of Guggenheimer Coffee, you get the taste experience of over 80 years of roasting craftsmanship and a little taste of the region close to Naples, Italy where the magic happens. A great cappuccino or espresso is part of life's peak moments, as well as a daily ritual. At Guggenheimer Coffee they understand that to create this moment of indulgence, time is of the essence.

BRAND
This double-distilled Single Malt is produced in one of the worlds most northern distilleries

ESTABLISHED
1988

PRODUCTS
Scotch, Single Malt

WEB
www.malts.com

THE STORY
An unmistakable taste experience that took root in 1830, when the Talisker distillery was built by the brothers Hugh and Kenneth MacAskill in the shelter of the Cuillin mountains. The name Talisker comes from the old Norse "Thalas Gair". The distillery is situated on the 57th northern latitude, making it one of the northernmost whisky production sites in the world. This whisky is still produced by using traditional and well-tested methods.

The single malt is double distilled and unique with its full character and maritime depth. In addition to the unique location of the distillery, the choice of the barrels in which it is stored makes for some peerless notes. The taste is initially soft and sweet, with a slight tingling sensation on the tongue. The aroma fully develops with a hint of salted caramel and a slightly smoky note followed by gentle herbs, and then its finishing touch in the mild taste of the sea salt. The scent unfolds with a citrus-sweet note of fresh orange and is replaced by a touch of fresh seaweed, a woody note and sea salt freshness.

The Talisker Skye is all about the taste, which is achieved by choosing the right barrels, perfect timing and the Atlantic weather conditions. With Talisker Skye, Diageo, the owner company of Talisker, caters to both connoisseurs and whisky newcomers, who are looking for new taste experiences.

TALISKER

KLINGENBERGER SPÄTBURGUNDER

Benedikt Baltes bought the city of Klingenberg's winery in northern Franconia in 2010. His plan was to subsequently raise the varietal ratio on the red sandstone sites to 100% Spätburgunder. He has achieved this goal and his wine is about to be certified organic.

BODEGA CÀMBRICO 574 UVAS

One of the main ingredients in this wine is the Rufete grape, a native to the region around Salamanca. Bodega Cámbrico has made it their goal to re-cultivate this almost extinct variety. Set in altitudes between 700m and 900m, their vineyards produce full bodied, mineral wines.

FONTERENZA PETTI ROSSO

Petti Rosso is a fun, quaffable natural wine made mainly from Sangiovese. It expresses the grape in its pure form: fresh with a lively acidity and plenty of red berry on the palate. The grapes are harvested by hand and the wine unfiltered.

WEINGUT HOFFMAN RIESLING HUNDERTGULDEN

This full bodied Riesling is from Rhinehessen home to some of the most sought after dry Rieslings. Jürgen Hofmann has rediscovered the Hundertgulden vineyard which with its highly calcareous soils and old vines produces finely structured Rieslings.

AVAILABLE AT WALTER & SOHN WEINHANDLUNG

KOMETENWEIN
´ZIE*XV`DOLOMITI BIANCO

One of a limited-edition series of wines, the ZIE*XV comes from a wildly planted vineyard with over 30 different indigenous grape varieties from Greece, Southern France and other regions. Experimenting with innovative ideas, this brand is leading the path to sustainability.

MAS DES DAMES
LA DAME

Mas des Dames translates to 'farm of the ladies'. Dutch woman Lideweji van Wilgen has been running this organic winery together with her three daughters since moving to the Languedoc in 2002. The La Dame is a dark blend of Grenache, Syrah and Carignan, with spicy notes.

CHAMPAGNE
PLATINE 1ER CRU BRUT

The Platine is a blend of 80% Pinot Noir and 20% Chardonnay. Maturing in the bottle for 24 months before being released, a share of the reserve wines is aged in oak barrels enabling it not only to display delicate aromas of ripe fruit and hazelnut but also a toastiness.

GRÜNER VELTLINER
EHRENFELS

This fine example of Grüner Veltliner comes from the Kremstal next to the Wachau region. In 2011, the Ehrenfels vineyard was declared '1ste Lage', meaning one of the premier Cru sites of the region.

WWW.WALTERUNDSOHN.DE

ZACAPA

BRAND
This premium rum is aged in old bourbon and sherry casks, high up in the Guatemalan mountains

ESTABLISHED
1930

PRODUCTS
Rum

WEB
www.zacaparum.com

THE STORY
Zacapa is the most exquisite tasting rum in the world. It is an award-winning, premium rum from the mystical land of Guatemala and is recognized by the World Luxury Association as one of the 'World's Top 100 Most Valuable Luxury Brands'.

Zacapa is produced in the verdant Guatemalan mountains, and at 2,300m above sea level, this is one of the highest ageing facilities in the world. The key ingredient is virgin sugar cane honey which combined with the unique 'Sistema Solera' distillation process results in an exceptional rum with a distinct and remarkable taste.

Master blender Lorena Vasquez carefully combines ingredients into previously used whiskey and sherry casks, and at this high altitude and constant average temperature of 114°C (58°F) the rum retains its distinctive deep, rich flavors and qualities.

For Zacapa, time matters more than anything. It is a focus throughout every stage of the distilling process, from blending spirits of different ages and personalities and storing the liquid in selected barrels during the maturing process, right up to taking the time to sip, savor and enjoy the liquid – this is called the Art of Slow.

Ron
Zacapa
SISTEMA
23
SOLERA
GRAN RESERVA

Zacapa

WHY GOD GAVE US SUNDAYS

STEFAN BRAUN

ON ANY GIVEN SUNDAY IN MONOPOLI

Nothing cures the soul, wrecked from a hard week, better than a meal in good company. It creates a heart-warming connection that catapults our inner spirit to new heights. The most important ingredient for these types of meals is quite often the time spent in preparation by all the participants. An investment, if you will, in good faith on the wonderful outcome destined for the senses and the belly. When we sat in the office with photographer Stefan Braun to discuss layouts for a new book project, it was close to lunch. Opening the subject with where to eat, Stefan raved about a trip he had just come back from: A week in Monopoli, a small coastal town between Bari and Brindisi on the southern Adria coast. With the simple words, "You wanna see the pictures?", we were off on a visual adventure.

STEFAN BRAUN

STEFAN BRAUN

STEFAN BRAUN

STEFAN BRAUN

STEFAN BRAUN

STEFAN BRAUN

STEFAN BRAUN

STEFAN BRAUN

STEFAN BRAUN

CHAPTER FIVE

SERVE & CELEBRATE

We are a social animal. Without community, life for us can become disjointed. Coming together in ways that affirm life, our most important acts, such as eating and imbibing, enforce a solidity in the nature of who and what we are. We revel in these rituals to confirm our substance.

ANZA

BRAND
Espresso machine mede of concrete from Berkeley, California

ESTABLISHED
2017

PRODUCTS
Espresso Machines

WEB
www.anzacoffee.com

THE STORY

The coffee world has undergone rapid change over the past few decades. But in spite of industry innovations, little change has occurred around the aesthetics of espresso machines. Why is that? This question burned in the back of the minds of the design studio Montaag, who set out on a process to create something new. The result is the AnZa espresso machine, produced in two variations: one, brutalist concrete, and the other, an elegant white Corian.

The AnZa was created in a backyard shed by Montaag designers in Berkeley, California, using materials seldom found in kitchen appliances, not to mention espresso machines. The result is a spectacular espresso machine that will bring life to any kitchen, and an unparalleled conversation piece. After all, making espresso is an art, and the AnZa delivers beautiful coffee while working as a decorative sculpture.

The team will release the new production machines after four years of testing and refinement. The AnZa machines offer more than a strong cup of coffee and distinctive profile. Their concrete and Corian models will be equipped with connectivity, giving users remote access, IoT integration, and the ability to program machines to fuel espresso habits.

AnZa is available via the crowdfunding platform Kickstarter, with machines shipping at the beginning of 2018.

BETONART AMBIENTE

BRAND
With special casting methods, betonart-ambiente turns raw concrete into lights and bowls

ESTABLISHED
2015

PRODUCTS
Handcrafted concrete light and deco bowls

WEB
www.betonart-ambiente.de

THE STORY
Betonart-ambiente produces unique lighting and decorative bowls that are handmade using pure concrete and a special casting technology. The German founders, Carolin and Jörg Wähdel from Heidelberg, discovered their love of this craft. “We fell in love with this raw material, and we are thrilled by the uniqueness and personality it produces.” With the help of a complex and lengthy finishing process, Betonart-ambiente transforms the raw concrete bowls into fascinating designs that create a magical and warm atmosphere at any dining table. Each piece is a one-of-a-kind creation, and tells its own special story, so it’s not surprising that each piece has its own name, given according to the diversity of color, structure and surface. Made to be unique!

BLAUDRUCK KÓO

BRAND
Joseph Koó is one of the last surviving original Indigo dyers

ESTABLISHED
1921

PRODUCTS
Tablecloths, sheets, aprons, bags, shirts, scarfs and shawls

WEB
www.originalblaudruck.at

THE STORY

Once upon a time, indigo-blue print materials were the everyday workwear in all of Burgenland. Today, Blaudruck Koó is one of the last original workshops in Europe, printing these clothes and other products with the old technology and materials, dyeing with real indigo. Founded in 1921 by Joseph Koó's grandfather, now he and his wife run the business together. Nearly 100 years old and run by three generations, the business has rarely been closed for even one day. Although tradition is important, staying current and experimenting with new things is critical.

To call it "blue print" is actually misleading, because the print is made by the absence of that color. With wooden molds or rollers (some over 200 years old), the Papp (a mixture of gum arabic and alumina) is applied to the white cloth. The cloth is then dipped into an often decades old indigo dye. The result is a white pattern on a blue background. After printing, the material must dry for about 4 - 6 weeks, before it again goes through the dyeing process. This process can vary depending on the desired intensity of the blue tone and last up to four hours. Afterwards, the Papp is washed out and the material is dried outdoors. Their specialty is double prints, having a different pattern on each side of the product.

In 2010, the Burgenland Indigo hand-pressed "blue print" was listed as a Cultural Heritage Technique by the expert advisory board at UNESCO.

BEAR POND ESPRESSO

BRAND

Bear Pond Espresso is a Japanese Mecca for coffee lovers from all around the globe

ESTABLISHED

2009

PRODUCTS

Coffee, espresso

WEB

www.bear-pond.com

THE STORY

Katsu Tanaka is owner and barista of Bear Pond Espresso. He is also midwife to a special liquid creation he calls “Angel Stain”. It is a bold statement in fluid form. Confidently expressing that espresso is not only taste, it is a feeling, and when you drink it, you definitely discover new feelings, about your future potential, about your own capabilities, about life.

In 2009, Katsu was one of the early coffeepreneurs, and one of the first in Tokyo, Japan to offer his customers the experience of a “coffee stand with its own roaster”. This new style has since become very popular worldwide.

All of BEAR POND ESPRESSO coffees are harvested at the peak of their flavor and their roasting techniques are designed to enhance the location, elevations, soils and climate that give each coffee its unique individuality. These quality espressos are brewed to perfection with the extraction arts of highly trained baristas.

One of their most popular offerings is "Flower Child", an espresso blend with a subtle flavor body that is at once spicy, salty, and silky, mingled with a suggestion of dark chocolate.

Bear Pond
ESPRESSO
ROASTERS

Flower Child
BEARPOND
Espresso
ROASTERS

BEAR POND
Espresso Roasters

BEAR POND
Flower Child
Bear Pond
ESPRESSO

BULTHAUP

BRAND
Luxury products with an emphasis on design

ESTABLISHED
1949

PRODUCTS
Living space design, accessories, furniture

WEB
www.bulthaup.com

THE STORY

In 1949, Martin Bulthaup founded his company and two years later bulthaup began making kitchen furniture. By the late 1960s, the company was already one of the top German kitchen manufacturers. In 1978, Martin's son Gerd Bulthaup took over and came into contact with the designer Otl Aicher, whose ideas were largely based on simplified forms and functionality. This set the tone for bulthaup's creations, going forward.

With bulthaup's launch of b Solitaire, there is now an additional focus for these Makers. A path rooted in the kitchen but branching out to the whole living space. Filling the environment where we spend most of our time with products known for their functional and ergonomic designs.

bulthaup knows that the heart of any kitchen is the table. It is the focal point of the living space, a venue for all of its denizens highs and lows. People spend valuable time with their loved ones here, enjoying food and drink and, most importantly, each other. The essence of bulthaup design is to offer furniture that invites people to spend precious moments together. Now the bulthaup b Solitaire table, bench and bench cushion in leather accompany these extraordinary moments.

All the b Solitaire elements can be customized, for example by choosing the surface structure and material, its height and generally its size and the way it is equipped.

bulthaup b Solitaire represents the entire living space. b Solitaire gives space for flexibility, but can also be integrated as familiar elements. They accompany and grow with people along their journey throughout life. Humans feel most at ease in the company of others and when they are surrounded by personal items that are important to them. bulthaup b Solitaire brings ways to do this directly into all facets of the living space.

bulthaup

bulthaup

FRITZ MÜLLER

BRAND
German answer to Prosecco that stirs up wine menus and dusty cellars

ESTABLISHED
2010

PRODUCTS
Sparkling Wine

WEB
www.fritzmueller.fm

THE STORY
A massive success story since 2010, Fritz Müller has what German sparkling wines have been sorely lacking: loads of charisma and obvious sex appeal. Often called Germany's answer to Prosecco, Fritz Müller charms with its flamboyant packaging and funky labels, but there is more to these wines than meets the eye. Produced from the abundantly planted Müller-Thurgau varietal, which creates fresh, natural wines suitable for a euphoric fizz, they are the brainchild of two young German wine professionals, Guido Walter and Jürgen Hofmann. Guido is the man behind the idea and the concept, while Jürgen makes these exceptionally crafted wines on his family's property in Rheinhessen.

The regular Fritz Müller Perlwein is a perfectly clean and balanced wine, with mild, appetizing bubbles. The makers unabashedly encourage consumers to use the Perlwein as a basis for a fizzy cocktail, for which it is perfectly suited due to its freshness and neutrality. One step up, is Fritz's big brother, Der Dicke Fritz, which is produced from 35-year old vines, using the 'Méthode Rurale', a traditional method rarely used in Germany. Extremely fruity, with hints of ripe apple, pear and grapefruit, this outstanding wine is best served lightly chilled, in a champagne glass. Fritz Müller's naturally fruity and vibrant offerings set them distinctly above the crowd.

FRITZ MÜLLER
PERLWEIN
MÜLLER-THURGAU
TROCKEN

METHODE RURALE
DER DICKE FRITZ
BRUT ZERO

IITTALA

BRAND
An icon of Finnish design, Iittala has long been synonymous with extraordinary glassware

ESTABLISHED
1881

PRODUCTS
Vases, glassware, glass making

WEB
www.iittala.com

THE STORY

What started as a glass factory in Iittala, Finland, today celebrates generations of essential objects that are made to enrich people's everyday lives. The belief that objects should be distinctive, combinable and multi-functional, with a lasting design that inspires individual use and expression is the very cornerstone of the culture in this shop. The progressive philosophy of design heroes like Kaj Franck and Alvar Aalto continue to inspire Iittala to keep itself forever relevant. They don't just create beautiful objects, they believe in timeless design that will never be thrown away.

Looking like nothing else, the Alvar Aalto Collection laid a foundation for this timeless design template. And since then, it has become an icon of Finnish design and one of the world's most famous glass objects. Craftsmanship forms the basis for mouth blown Alvar Aalto vases, designed by internationally renowned architect Alvar Aalto in 1936. It takes 7 craftsmen, 12 work stages, 1100 °C and 30 hours to make a single Aalto vase. The skillful craftsmen of the Iittala glass factory have dedicated themselves to keeping this tradition alive and every single Aalto vase is individually made.

BRAND
Design aesthetics connected to the bauhaus, Mono's cuttlery has become a favorite among design-minded diners

ESTABLISHED
1959

PRODUCTS
Flatware, cutlery

WEB
www.mono.de

THE STORY
In 1895, this fifth-generation family business was founded and the brand Mono was established in 1959 with the production of Mono-a flatware. Designed by Peter Raacke, the Mono-a flowed cleanly and to the point, its geometric form, made of industrial steel. In the early 80s, Designer Tassilo von Grolman designed a handmade glass vessel and an oversized sieve, stating that this is what tomorrow's teapot would look like. With the clean Bauhaus influenced design the canister did, however, reflect an aesthetic design extension of Mono-a. The story of the Pott Edition began in 1904 when Carl Hugo Pott opened a factory for the honing of cutting tools. His son, Carl Pott, went into the business in 1932 and formed a flatware business that continues to enhance the reputation of the Pott brand. In over half a century, he generated an iconic flatware program that has been awarded a myriad of prizes. In 2006, the Seibel family, with the Makers of Mono, decided to maintain that historic product portfolio, and continue to produce these timeless classics. Since 1959, Mono's goal has been to develop designer products that follow no trend, but simply outlast all. Striving to produce items that do not decorate but function and have the highest standards of production, material and originality.

MONO

GS
TAF
2784
NBS
12 12 88
T
VORL. 784
E Z H
24.11.54

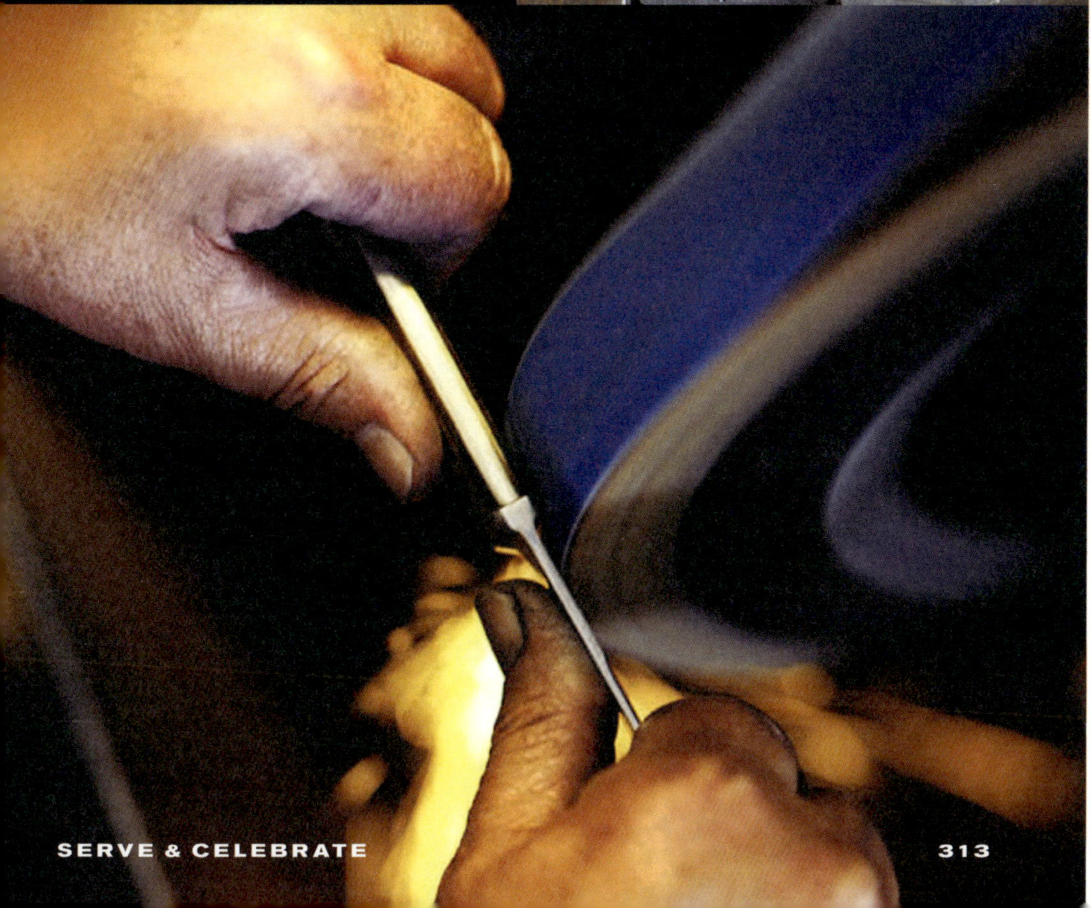

REICHENBACH

BRAND
Tradition and innovation are joined seamlessly at Reichenbach's porcelain factory

ESTABLISHED
1900

PRODUCTS
fine porcelain for daily life and special occasions

WEB
www.porzellanmanufaktur.net

THE STORY

In 1900, the Porzellanmanufaktur Reichenbach was founded when nine local porcelain painters got together and laid the foundation. The type and quality of handwork that still characterizes production today, is expressed in the company motto: Traditional art meets modern design.

Products consist of high quality raw materials, refined with modern production techniques by well trained and experienced craftsmen. These carefully selected materials, with an eco-friendly mindset, form the basis of quality products in ranges that have increased since the founding of the company.

Customers can order one special piece or a set of pieces for household, hotel, or decoration. Skilled workers with an eye for detail assure consistency. Tradition, innovation and design-oriented product development are integrated into every part of the making of this fine china.

This is the home of excellent and exquisite collections that are "100% Made in Germany". Being able to use this seal helps Reichenbach stand out from the crowd. In addition, Reichenbach has been closely associated with Thuringia porcelain since 1830 and products at Reichenbach have been designed by renowned artists such as Gerd Sommerlade, Paola Navone, Metz and Kindler, Blohmgumm, ServiceForm, and Christine Klauder.

ROCKET ESPRESSO

BRAND
Premium domestic espresso machines replicating the performance of professional machines

ESTABLISHED
2007

PRODUCTS
Espresso machines

WEB
www.rocket-espresso.com

THE STORY
New Zealander Andrew Meo and Italian Daniele Berenbruch got together in Milan, Italy with a clear vision to produce the best possible domestic espresso machines.

The philosophy was to take the technology used in larger commercial machines and incorporate it into a smaller machine. A machine that is designed to produce outstanding espresso, equal to that served in a café, but in the comfort of your own home.

To set the brand apart from other machines offered in the market, the focus at Rocket Espresso is on producing beautiful and contemporary styled machines, with an almost obsessive attention to detail.

Handmade in their Milan factory, it's unusual today to find a product that is built to last 20 to 30 years if properly maintained.

The Rocket Espresso brand has evolved further, now adding commercial espresso machines into their portfolio, which can be found in some of the best cafés in the world.

An obsession with the perfect espresso has led to the opening of their own coffee roastery, Salva Vita, which means 'life saver'.

A New Zealander and an Italian bringing together unique ideas from the opposing ends of the world, to achieve the common goal of bringing the best possible espresso to your cup.

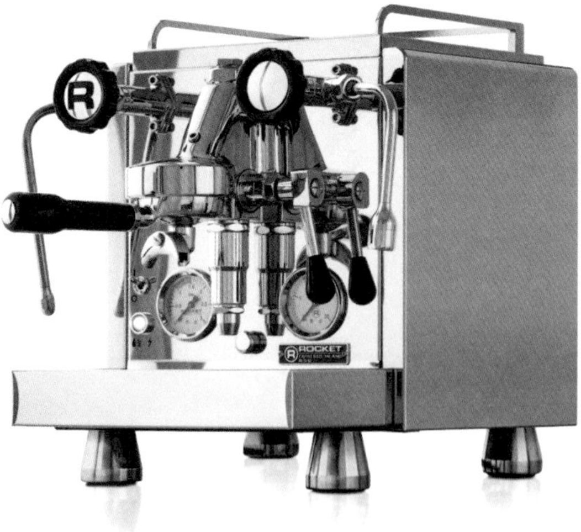

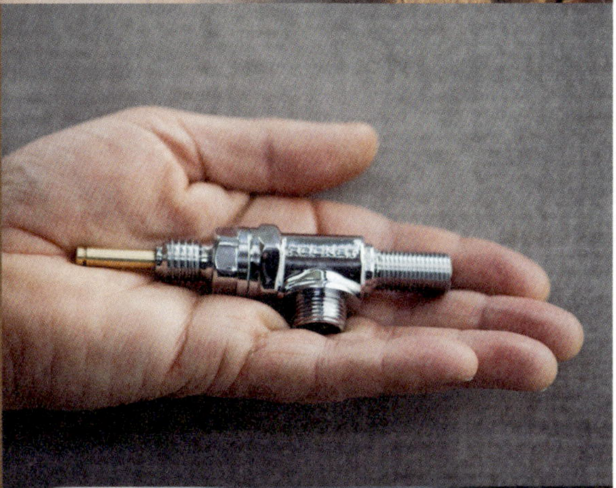

SAWADE

BRAND
Berlin's oldest chocolate factory offers handmade delights in all flavors, shapes and quantities

ESTABLISHED
1880

PRODUCTS
Fine chocolates and pralines, bonbons, truffles

WEB
www.sawade.berlin

THE STORY
Sawade is Berlin's oldest chocolate factory which makes pralines, confections, and bonbons. The founder Ladislaus Ziemkiewicz named his business after a good friend and neighbor Madame Marie de Savade after a visit to Paris which inspired him to enter the chocolate business. The chocolatier was eventually nominated by Prince George of Preussen and made a royal supplier of chocolates, which was a very high honor.

In 1970, there was a takeover which placed Sawade in the Reinickendorf section of Berlin where it is still located. In 2013, new owners came along, a married couple named Hübel, who maintain the tradition of handmade chocolates and the first class quality represented by the Sawade name. Five new stores openend up in Berlin with a new design. Chocolates can now be shipped worldwide thru the online shop.

Chocolates can be bought singly, in boxes, or as gifts. With or without alcohol, all quality ingredients, the chocolates are as unique as the clientele.

"It's our passion to make exquisite, handmade chocolate products known to the world again. What people once treasured is again in demand today. Quality was, is, and always will be a pleasure."

Sawade

HERING BERLIN

BRAND
Minimal and modern shapes give Stefanie Hering's tableware a fresh spin

ESTABLISHED
1999

PRODUCTS
Fine porcelain, matte bisque porcelain

WEB
www.heringberlin.com

THE STORY
As a young girl, Stefanie Hering's dream was to be a porcelain maker, so she just started creating things on a potter's wheel, letting her imagination soar. Now Stefanie, founder and CEO of Hering Berlin, is the most innovative and modern porcelain manufacturer in the capital city of Berlin. She creates products that ordinary folks and top chefs both rave about, sensual bisque porcelain for daily use and every occasion, extremely durable, eco-friendly and "hygienic to the max" as Stefanie says.

The unglazed collection with evocative names such as Cielo, Velvet, and Pulse delight customers with their velvety-soft surfaces. Matte bisque porcelain is the unmistakable hallmark of Hering Berlin, which has become a leader in porcelain manufacturing.

At its core, Hering Berlin has a simple recipe for success: create traditionally hand-crafted, elegant and contemporary tableware for every day. Their master porcelain makers employ elaborate craftsmanship at the factory in Reichenbach to create unique and exquisite bisque porcelain, by integrating techniques into sophisticated designs, creating timeless artwork.

SWISS APÉRO

BRAND
High-quality, gourmet meat specialities for an exclusive Apéro

ESTABLISHED
2015

PRODUCTS
Air-dried meat specialities

WEB
www.swiss-apero.com

THE STORY
In these times of higher, faster, further – particularly in our modern age, little moments of calmness are very important. During a traditional apéro, fine, air-dried meat specialties such as the famous Bündnerfleisch are arranged with aromatic cheese on a slate or plate. Accompanied by a glass of wine and good conversations, relaxation comes easily. Stress and rush are a thing of the past. The apéro, a tradition deeply embedded within Swiss culture, is the perfect timeout.

In the mountain regions, meat has been made durable through air-drying since the 14th century. Once a staple food for mountain farmers, now a unique delicacy and one of the most precious products in the world. Over the centuries, the production has barely changed: Meat of the highest quality, clear mountain air and a lot of time are the secret to these specialties.

This serenity can still be felt and tasted today. Due to the long aging process in mountain air, the meat gets a unique taste, which unfolds perfectly when cut into fine slices. Anyone wondering what might be the best time for an apéro, should go about it in a relaxed manner. Whether with a gathering of guests, at a birthday party or during an evening for two, you may enjoy a Swiss Apéro on any occasion and prove that even the little moments of everyday life can be special.

SYMPOSIARCH

BRAND
Jane Fenzel's furniture design project playfully incorporates the social components of a meal with friends

ESTABLISHED
2017

PRODUCTS
Dinnerware and dining tables

WEB
www.janefrenzel.de

THE STORY

A symposiarch, by definition, was a person who would manage the flow of dishes and conversation during a feast in ancient Greece. He would make sure the ease of conversing was not hampered by the serving of meals or placement of chairs and tables. At Symposiarch, a smart table has been developed that now performs all these functions.

The layout of the smart table is concentric rings that start off expanded to allow for the main course and progressively fold into each other as the meal ends in dessert, keeping the diners as close together as possible and still allotting room for the required dinnerware.

The dinnerware is also sized depending on which course is served. The main course is served on individual plates, whereas dessert is served on one plate that everyone shares. It's a playfully orchestrated table dance to keep the diners engaged in conversation with each other.

WERKSTATT:MÜNCHEN

BRAND
Individual works of art that also serve as tableware

ESTABLISHED
1998

PRODUCTS
Real silverware, silver dinnerware

WEB
www.werkstatt-muenchen.com

THE STORY
A place long synonymous with artisanal beauty and industrial romanticism, Werkstatt: München was founded in 1996 by Klaus Lohmeyer. The brand features jewelry, lifestyle objects and leather goods, entirely hand made in the atelier in München, Germany.

WERKSTATT:MÜNCHEN blends together meaning, function and timeless elegance into finely crafted pieces. Drawing inspiration from contemporary culture, past memories and symbolism, Klaus aims to embellish people with their own personality. The industrial vibes, that shape Werkstatt:München's creations, emerge from the company's home, an old foundry located in the heart of Munich.

The small irregularities and vivid surfaces make each design truly unique by respecting the natural reactions of metals to the working process. With rough, genuine creativity and solid skills, the timeless value of precious things to be inherited, passing on stories and traditions from generation to generation are re-established.

WERKSTATT:MÜNCHEN's vision of tableware reflects their unique approach to "the unfinished" in perfection. In this state of the process, the rough appearing objects show the maximum of power, labor and precision worked into them through traces of the handmolding process. You can find marks of oxidation, caused by the torch to heaten for recrystallization, mixed with blank silver spots, exclusively generated by a silversmith forging hammer and an iron anvil. All pieces have their individual surface, evolved by the process of slowly and strongly moulding stroke upon stroke. Within the workshop's suggestive atmosphere, Klaus and a skilled team labor with passion and precision to create pieces that have a soul.

LED ZEPPELIN

ZALTO GLAS

BRAND
Fine glassware from Austria's Waldviertel Region that impresses enologists around the world

ESTABLISHED
2006

PRODUCTS
Wine glasses, beer glasses, decanters, glassware

WEB
www.zaltoglas.at

THE STORY
The Zalto brand has been created and groomed by the Zalto family, a glass dynasty from Veneto. Six generations ago the family settled in the northern Waldviertel and began producing, in Neunagelberg, a high-quality mouth-blown glass, never before seen. Re-established in Neunagelberg in 2006, Zalto began to specialize in superior, mouth-blown jars with the acquired brands of Zalto and Denk`Art.

Zalto Glas has been consistently committed to the tradition of offering wine glasses and other libation glasses to connoisseurs in all of their unusual forms. All Zalto drinking glasses are hand-blown at the highest level of craftsmanship.

The glasses of Zalto not only impress with timeless designs but are all made without the addition of lead oxide and are resistant to turbidity. Despite its thin-walled structure, the Denk`Art series fulfills all the attributes of a modern glass with regard to care and longevity. Denk`Art glasses are dishwasher-safe and have a high level of practicality.

A Zalto glass does not come between the drinker and what is drunk.

APPENDIX

FIND ALL OF THESE BRANDS IN THE STORE FINDER ON:

www.makersbible.com

Cut, Cook & Burn

DUALIT
www.dualit.com
156

Preserve & Season

EDELKONSERVEN
www.edelkonserven.at
209

Cut, Cook & Burn

EPICUREAN USA
www.epicureancs.com
158

Preserve & Season

EXALTA
www.exaltachocolate.com
210

Hydrate & Distill

FENTIMANS
www.fentimans.com
258

Cut, Cook & Burn

FEUERRING
www.feuerring.ch
160

Cut, Cook & Burn

FLORENTINE KITCHEN KNIVES
www.florentinekitchenknives.com
164

Hydrate & Distill

FREIMEISTER
www.freimeisterkollektiv.de
260

Cut, Cook & Burn

FRIEDR. DICK
www.dick.de
166

Serve & Celebrate

FRITZ MÜLLER
www.fritzmueller.fm
308

Hydrate & Distill

HEILAND
www.heiland-liqueur.de
217

Provide & Cultivate

HELDBERGS
www.heldbergs.com
104

Serve & Celebrate

STEFANIE HERING BERLIN
www.heringberlin.com
320

Provide & Cultivate

HOFBRÄUHAUS-KUNSTMÜHLE
www.hb-kunstmuehle.de
99

Cut, Cook & Burn

HOFFMANN-METALLGEFÄSSE
www.hoffmann-metallgefaesse.de
170

Hydrate & Distill

HOPFENGUT 20
www.hopfengut.de
262

Provide & Cultivate

HUNTER
www.hunterboots.com
106

Serve & Celebrate

IITALA
www.iittala.com
310

Hydrate & Distill

ISAR BIER
www.isar-bier.de
265

Provide & Cultivate

JECKYBENG
www.jeckybeng.com
89

Preserve & Season

KNALLE POPKORNDITOREI
www.knalle.berlin
218

Cut, Cook & Burn

KUPFERMANUFAKTUR
www.kupfermanufaktur.com
172

Preserve & Season

LES RHUMS DE CED
www.lesrhumsdeced.fr
220

Preserve & Season

MEIN MÜHLENFELD
www.meinmuehlenfeld.de
219

Hydrate & Distill

MEZCAL ATENCO
www.mezcal-atenco.com
266

Cut, Cook & Burn

MOCCAMASTER
www.moccamaster.nl
174

Serve & Celebrate

MONO
www.mono.de
312

Provide & Cultivate

MORITZ WENZ
www.moritz-wenz.de
110

Provide & Cultivate

MOORMANN
www.moormann.de
108

Hydrate & Distill

NARDINI
www.nardini.it
268

Provide & Cultivate

WALDKAUZ
www.waldkauz.net
126

Hydrate & Distill

WALTER & SOHN
www.walterundsohn.de
282

Serve & Celebrate

WERKSTATT:MÜNCHEN
www.werkstatt-muenchen.com
326

Cut, Cook & Burn

YRON
www.yron.de
194

Hydrate & Distill

ZACAPA RUM
www.zacaparum.com
284

Serve & Celebrate

ZALTO GLAS
www.zaltoglas.at
328

GLOSSARY

GLOSSARY

THE MAKERS BIBLE TEAM

Philipp Esser, Sara El Abed, Johannes König, Florian Brugger, Michael Schmidt, Yumi Kimoto, Lars Harmsen, Gert Doering

We are a multidisciplinary team of different generations and backgrounds that all have one thing in common: our love of the authentic, the built-to-last, the products and brands created with an almost extinct craftsmanship, made with soul. United in purpose previously, with our first book becoming a rousing success, it was the feedback from readers, traders and makers that encouraged us to go even further.

Pushing the envelope to foods, drinks and kitchen utensils, we re-assembled on our new quest to pull together a further brand compendium and share it. Our goal, as always, is to build a lasting platform, a clear guide for people to discover brands that represent these inalienable values.

OUR CONTRIBUTORS

HEIMO TSCHERNE

MÜNCHEN (D)

A networker by profession Heimo is a specialist on food and drinks' brands and communicating their needs. "Food-PR" is one way of calling it; when talking to Heimo, it is obvious that this food & beverage expert from The Steiermark has made a profession out of his passion.

BOTHO STEIN

HEINSBERG (D)

As professional chef Botho provided valuable input for this book. Formerly pursuing a career on various posts in various star-spangled restaurants. Among them Kieler Kaufmann, Auberge des Glazicks and L'Atlantide, both in Brittany. He now works as an exclusive event chef in front of people, enjoying the contact and their appreciation, rather than in a remote kitchen separated from the guests.

ROBERT WEBER
WIEN (AT)

Actually an advisor in branding, marketing and communication Robert is a Mad Man in the kitchen. But he is more than a dedicated hobby chef. Not only did he sort out the Austrian food market for us but teaches cooking to friends, too.

MORITZ FUCHS
MÜNCHEN (D)

Founder and chief organizer of New Heritage, a makers' fair for true and honest products and brands, Moritz and his team have successfully created their fair out of nothing by networking and reliable delivery to their promises.

RUSS VIOLA
COCOA BEACH (USA)

Born in the swamps of Louisiana, raised in the woods of northern Massachusetts, living on a beach in Florida Russ is author, copywriter and storyteller. His wisdom of words, his sharp mind plus his Swiss-clockwork-like reliable rhythm makes him a valuable contributor and in this case, too a curator of this book.

DOERTE WELTI
BINZ (CH)

Doerte Welti, German born, lives in Switzerland. She has studied Art History, Architecture and Marketing. A book on quality products without Swiss participation is hardly achieveable. Doerte has combed the Swiss market for us. Easy for her, as a journalist on travel, food and life-style, a PR-manager and copywriter she gave valuable input into the curation.

MAKERS BIBLE

PUBLISHER
Melville Brand Design GmbH
Goethestr. 20, 80336 Munich
Germany
www.melvillebranddesign.com

This book was conceived, edited and designed by Melville Brand Design

EDITORIAL
Editors – Gert Doering, Lars Harmsen, Michael Schmidt
Project Manager / Main curator – Michael Schmidt
Product Manager – Gert Doering
Copy / Editing – Russ Viola
Creative Director – Lars Harmsen
Design / Layout – Dominic Brighton, Florian Brugger, Philipp Esser, Johannes König
Illustrations – Dominic Brighton, Lars Harmsen, Johannes König
Photography / Editorials – Philipp Esser, Lars Harmsen, Johannes König, Michael Schmidt
Photography / Brands – Dominic Brighton, Philipp Esser, Johannes König
Retouching – Fabian Kern, Patrick Kern www.bildergut.de
Social Media – Gert Doering
Press – Heimo Tscherne www.undtscherne.com

STORE FINDER WWW.MAKERSBIBLE.COM
Johannes König, Kolja Buscher, Gert Doering

PHOTO CREDITS

Every effort has been made to identify and acknowledge the copyright holder of all pictures in this book. If not commissioned by Makers Bible, photographic material has been provided by the brands. The following are photo credits as given to us. Missing credits are unintentional. Copyright owners who have not been correctly acknowledged are invited to contact: info@makersbible.com

REPORTAGES
Cup Masters of the Universe – Horst A. Friedrichs www.horstfriedrichs.com
Men of Steel – Lars Harmsen www.melvilledesign.de
Spirit of Oaxaca – Fer Arce www.photograferarce.com
Why God gave us Sundays – Stefan Braun www.braunphotography.de

BRANDS & EDITORIALS
Daniel Gebhart de Koekkoek – Biohof Rohrauer
Barbara Bonisolli – Dahlmann Catering
Dennis Savini, Roger Harrison – Feuerring
Florentina Klampferer – Friedr. Dick
Trustin Timber – Frost River www.trustintimberproductions.com
Reinhard Lang – Hartls Öl
Fer Arce – Mezcal Atenco
Julia Rotter – Moorman/Custom Bus
Elena Heatherwick – Olia Hercules, Heritage on the Menu
Anna Dabrowski – Ölwerk
Marty Steinhausen – Otto Gourmet
Stefan Goetschl – Schmiedewerk Asbeck
Anthony Pérez – Schmidt&Briongos, Bee hive Yourself
Mònica Figueras – Schmidt&Briongos, Bee hive Yourself
Arnoud Overbeeke, Richard Hanson – Sneeboer
Marcella Breugl – Symposiarch
Leo Sanzo – Teochef, Praise the Lard
Stefano Scatà – Teochef, Praise the Lard

PRODUCTION

PRINT
Druckerei Vogl GmbH & Co KG
Georg-Wimmer-Ring 9, 85604 Zorneding
www.druckerei-vogl.de

BOOK BINDING
Conzella, Pfarrkirchen

MATERIAL
Cover material – Winter Company Napura Canvas
Book end paper – IGEPA Caribic
Inside paper – IGEPA Salzer touch 120

COPYRIGHT

Published by Melville Brand Design, 2017
ISBN: 978-3-00-057983-7

ACKNOWLEDGMENTS

Leonardo Nave and Mario Bonamigo for networking & hospitality, Levinia Manfrotto from Libreria Palazzo Roberti, Fontana Brothers Valter and Andrea and her Mama, Berthold Hoffmann und Eisengiesserei Max Ginter Gmbh & Co KG for 1455°C, Marianna D'Incà Levis from ProCaffé, Giada Bionda from La Marzocco both for insights on Italian coffee, Daniel Dobrogeanu und Johann Paintmeier for their passion, Jürgen Kernegger and Stefanie Brix from Winter Company, Sabine Reister from Igepa for the paper and Lorna Kavanagh for being first.

Alex Messenger, Beatrix Neulinger, Carina Schüssler, Cindy Westermeier, Erik Fülöp, Eva Rosenberger, Jana Durchleuchter, Kathrin Mühlhofer, Lilit Chatschatrian, Lisette Schönitz, Margit Spitzbart, Marie Bund, Martin Opercan, Moritz Paetz-Borchardt, Ralf Vogl, Ruben Briongos, Sara El Abed, Stefan Neubrand, Victoria Schmidt, Vicky Hauer-Bain, Yumi Kimoto.

Curative support: Christian Klemenz from Bierothek (Bamberg), Guido Walter von Walter & Sohn (München), Horst A. Friedrichs (London), Moritz Wenz (Düsseldorf), Ralf Fischer from Ralf's Fine Garments (München), Timo Thurner from Aqua di Monaco (München), Theo Ligthart from Freimeister Kollektiv (Berlin), Zanus Matteo "TeoChef" (Bassano del Grappa).

Our special thank goes again to our American copywriter Russ Viola in Florida who had to hide from two hurricanes this season.

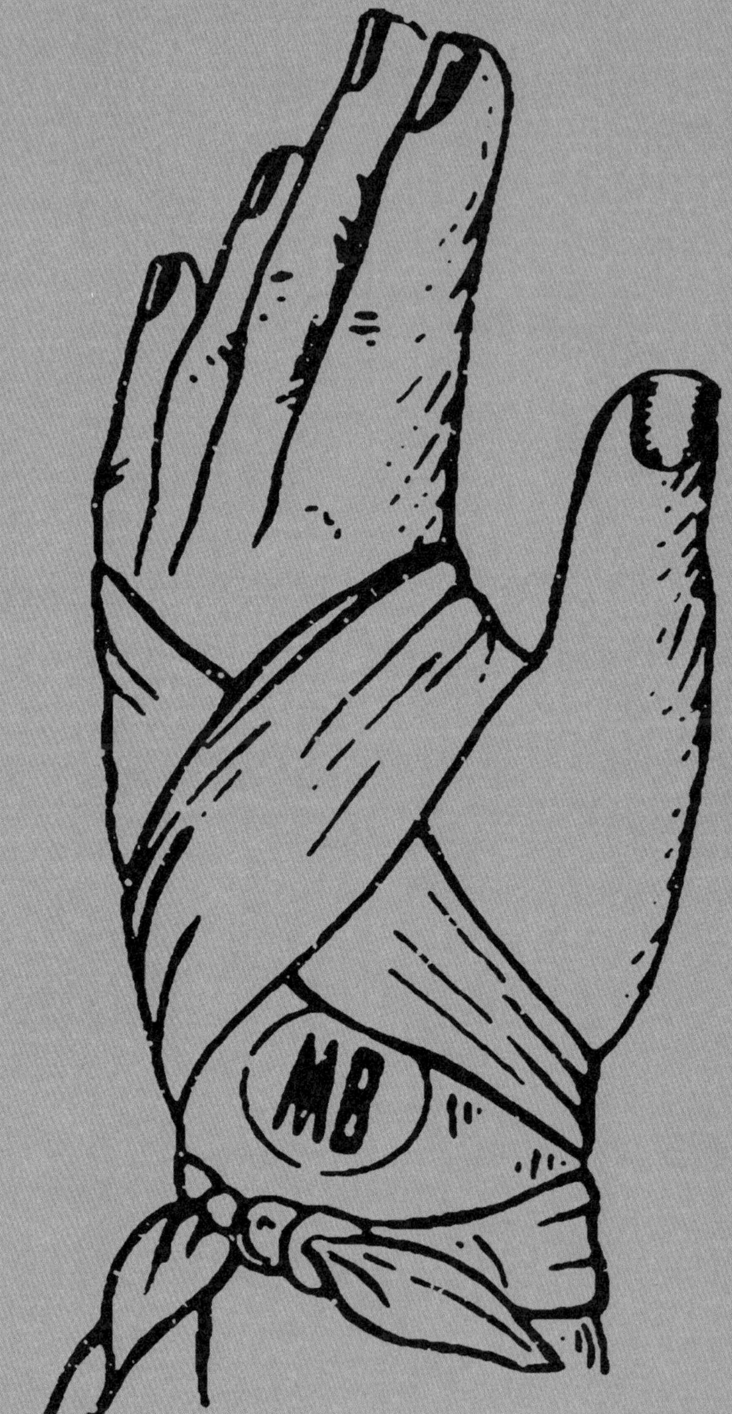
MB